T0094314

Design & Analysis of Clinical Trials for Economic Evaluation & Reimbursement

An Applied Approach Using SAS & STATA

Chapman & Hall/CRC Biostatistics Series

Editor-in-Chief

Shein-Chung Chow, Ph.D., Professor, Department of Biostatistics and Bioinformatics, Duke University School of Medicine, Durham, North Carolina

Series Editors

Byron Jones, Biometrical Fellow, Statistical Methodology, Integrated Information Sciences, Novartis Pharma AG, Basel, Switzerland

Jen-pei Liu, Professor, Division of Biometry, Department of Agronomy, National Taiwan University, Taipei, Taiwan

Karl E. Peace, Georgia Cancer Coalition, Distinguished Cancer Scholar, Senior Research Scientist and Professor of Biostatistics, Jiann-Ping Hsu College of Public Health, Georgia Southern University, Statesboro, Georgia

Bruce W. Turnbull, Professor, School of Operations Research and Industrial Engineering, Cornell University, Ithaca, New York

Published Titles

Adaptive Design Methods in Clinical Trials, Second Edition
Shein-Chung Chow and Mark Chang

Adaptive Designs for Sequential Treatment Allocation
Alessandro Baldi Antognini and Alessandra Giovagnoli

Adaptive Design Theory and Implementation Using SAS and R, Second Edition
Mark Chang

Advanced Bayesian Methods for Medical Test Accuracy
Lyle D. Broemeling

Advances in Clinical Trial Biostatistics
Nancy L. Geller

Applied Meta-Analysis with R
Ding-Geng (Din) Chen and Karl E. Peace

Basic Statistics and Pharmaceutical Statistical Applications, Second Edition
James E. De Muth

Bayesian Adaptive Methods for Clinical Trials
Scott M. Berry, Bradley P. Carlin, J. Jack Lee, and Peter Muller

Bayesian Analysis Made Simple: An Excel GUI for WinBUGS
Phil Woodward

Bayesian Methods for Measures of Agreement
Lyle D. Broemeling

Bayesian Methods for Repeated Measures
Lyle D. Broemeling

Bayesian Methods in Epidemiology
Lyle D. Broemeling

Bayesian Methods in Health Economics
Gianluca Baio

Bayesian Missing Data Problems: EM, Data Augmentation and Noniterative Computation
Ming T. Tan, Guo-Liang Tian, and Kai Wang Ng

Bayesian Modeling in Bioinformatics
Dipak K. Dey, Samiran Ghosh, and Bani K. Mallick

Benefit-Risk Assessment in Pharmaceutical Research and Development
Andreas Sashegyi, James Felli, and Rebecca Noel

Published Titles

Published Titles

Chapman & Hall/CRC Biostatistics Series

Design & Analysis of Clinical Trials for Economic Evaluation & Reimbursement

An Applied Approach Using SAS & STATA

Iftekhar Khan

University College London, UK

CRC Press
Taylor & Francis Group
Boca Raton London New York

CRC Press is an imprint of the
Taylor & Francis Group, an **informa** business

A CHAPMAN & HALL BOOK

CRC Press
Taylor & Francis Group
6000 Broken Sound Parkway NW, Suite 300
Boca Raton, FL 33487-2742

© 2016 by Taylor & Francis Group, LLC
CRC Press is an imprint of Taylor & Francis Group, an Informa business

Printed on acid-free paper
Version Date: 20150722

International Standard Book Number-13: 978-1-4665-0547-6 (Hardback)

Library of Congress Cataloging-in-Publication Data

Khan, Iftekhar.
   Design & analysis of clinical trials for economic evaluation & reimbursement : an applied approach using SAS & STATA / Iftekhar Khan.
     pages cm. -- (Chapman & Hall/CRC biostatistics series)
   Includes bibliographical references and index.
   ISBN 978-1-4665-0547-6 (alk. paper)
   1. Medical economics. 2. Clinical trials--Economic aspects. 3. Cost effectiveness.
   I. Title. II. Title: Design and analysis of clinical trials for economic evaluation and reimbursement.

R728.K43 2015
338.4'73621--dc23   2015021743

Visit the Taylor & Francis Web site at
http://www.taylorandfrancis.com

and the CRC Press Web site at
http://www.crcpress.com

These are our works, these works our souls display;
Behold our works when we have passed away.

First and foremost, I give thanks to God Almighty for giving me the health and energy to complete this project. I dedicate this book to those who I remember in my dreams and made this possible: my wife Suhailah, my children Yohanis and Hanzalah, my late father Ayub Khan and Asghari Khan, who was always kind to me and provided much encouragement. I also thank all those who supported me while I was completing this project, family and friends, past and present.

Contents

Preface

Economic evaluation is a firmly established discipline in the field of health research and policymaking. It has also become an increasingly essential component of clinical trial design to show that new treatments (healthcare technologies) offer value to the payers in various healthcare systems.

Several books already exist which address theoretical or practical aspects of cost-effectiveness analysis. What makes this book different from others is its insight into how health economic evaluation is applied in a clinical trial context in both academic and pharmaceutical/commercial settings. This book is not just about performing cost-effectiveness analyses, but it also emphasises the strategic importance of economic evaluation and offers guidance and advice on the complex factors at play before, during and after an economic evaluation has been performed. In addition, this book bridges the gap between applications of economic evaluation in industry (mainly pharmaceutical) and what students may learn in university courses.

An additional feature of this work is the availability of software, such as SAS and STATA code, for performing, in some cases, daunting computations. In addition, Windows-based software (SCET-VA® version 1.0) for sample size and value of information analysis, which is a menu-driven package avoiding the need for complex programming, is available free of charge. Therefore, this book will be a valuable source of knowledge for students who might pursue a career in this field or simply wish to know more about applying economic evaluation techniques.

Design & Analysis of Clinical Trials for Economic Evaluation & Reimbursement: An Applied Approach Using SAS & STATA is structured to cover many of the topics in economic evaluation that students, clinicians, health economists, epidemiologists, statisticians and other healthcare researchers will come across when designing a clinical trial for cost-effectiveness. No knowledge of economic evaluation, statistics or SAS is required to understand the key concepts in each chapter of the book, and the reader may ignore the SAS/STATA code and read interpretations of the output (the interpretations would be the same with any software).

This book can be used as either a self-study book, a textbook or a reference book. The format of the book is as follows.

Chapter 1 offers the reader who is unfamiliar with economic evaluation the economic background and context. Important economic concepts such as opportunity costs, discounting and the marketing authorisation landscape are reviewed. The clinical hypotheses are discussed in the context of economic hypotheses. The chapter discusses practical examples of strategic thinking necessary when developing a new treatment for market authorisation.

Chapter 2 introduces the incremental cost-effectiveness ratio, the cost-effectiveness plane and the different types of economic evaluation undertaken using the incremental net monetary benefit approach, and distinguishes between statistical and economic models. Chapter 3 elaborates on the detailed aspects of clinical trial design for cost-effectiveness, covering issues such as choice of comparator, timing of assessments, clinical report form design, treatment pathways and heterogeneity. The chapter contains a case study from a lung cancer trial.

Chapters 4 and 5 contain material on the analysis of resource use (costs) and quality of life, respectively. Chapter 4 discusses some of the more recent statistical issues and approaches to modelling cost data, distinguishing carefully between modelling health resource data and costs data. Chapter 5 covers the different measures of quality of life used in economic evaluations, such as EQ-5D, quality-adjusted life years, disability-adjusted life years and quality of time spent without symptoms of disease and toxicity. A section also includes more recent research into evaluating the sensitivity of generic measures to detect treatment effects. Mapping functions are also covered: methods applied in practice to estimate health utilities but not often covered in standard texts.

Chapters 6 and 7 detail the approaches to economic modelling and sensitivity analysis. Chapter 6 elaborates on decision trees and Markov and patient-level simulation approaches as well as discussing strategies for checking the Markov model – an area rarely discussed in standard texts but almost always required in industry applications. Chapter 7 provides details on approaches to simulating univariate and multivariate data for probabilistic sensitivity analysis. The complexities of multivariate simulation for probabilistic sensitivity analysis are, again, rarely discussed in most books on health economic evaluation.

Chapter 8 provides extensive details on sample size and power in a Bayesian and frequentist context. A demonstration of the freely available software for sample sizes and value of information methods using SCET-VA v1.0® discusses how to enter inputs to obtain the sample size required for a given probability of cost-effectiveness.

Chapter 9, although seemingly the odd one out, is still nevertheless relevant, since in most industry applications of cost-effectiveness and economic evaluation, evidence synthesis is often required. Indirect comparisons are sometimes needed to compare treatments that were not compared in the same trial, and some reimbursement agencies do request these. For this reason, this chapter was included, as it complements the methods applied in economic evaluation, particularly for market authorisation.

Chapter 10 is dedicated to examples of economic evaluation specific to cancer trials. Key design aspects of Phase III cancer trials, issues such as extrapolation of survival times after a study follow-up has been completed, treatment switching after disease progression and the relationship between transition probabilities and hazard ratios, as well as other practical considerations when designing and analysing cancer trials for economic evaluation, are discussed. Again, this is an area which is needed, but many standard texts do not distinguish between the peculiarities of economic evaluation in cancer trials and other therapeutic areas. Chapter 11 finally summarises the details and challenges of the reimbursement market in practice and further areas of research in the area.

I candidly admit that my objectives have been set high when structuring this book. Economic evaluation covers several disciplines, and addressing all of these has been challenging. I hope that the material in this book is suitable for a range of researchers of varying abilities, so that some will find the entire book useful, whereas for others some chapters will be useful. As a student textbook, it can be complemented by additional reading materials suggested in the bibliography to provide a book whereby students can be more prepared for the world of industry.

I acknowledge the anonymous reviewers of this book and the early discussions with various researchers at University College London, which have helped to improve the text. I am also grateful to the Cancer Research UK (CRUK) and Cancer Trials Centre, who respected my desire to see this project to completion while running multiple clinical trials. Finally, I thank Drs Suhailah Khan, Yohanis and Hanzalah for putting up with my countless hours on this book.

Iftekhar Khan
University College London
https://www.researchgate.net/profile/Iftekhar_Khan5

Author

Iftekhar Khan is a statistician, health economics researcher and academic at University College London (University of London). He has been an applied statistician for over 15 years in clinical trials, having worked in pharmaceutical companies and academic clinical trials units. Dr. Khan has earned degrees in statistics and mathematics from King's College London, the University of Kent and the University of Cambridge, including a master's in health economics and a PhD in health economic modelling (University College London).

Acronyms

AUC	area under the curve
BB	beta binomial
BNF	British National Formulary
BVN	bivariate normal distribution
CBA	cost–benefit analysis
CEA	cost-effectiveness analysis
CEAC	cost-effectiveness acceptability curve
CLAD	censored least absolute deviation
Cmax	maximum concentration
COPD	chronic obstructive pulmonary disease
CRF	case report form
CSM	condition-specific measures
CSR	clinical study report
CTAAC	Clinical Trials Awards Advisory Committee
CUA	cost–utility analysis
DALYs	disability-adjusted life years
DCF	data collection form
ECG	electrocardiogram
EE	economic evaluation
EMEA	European Medicines Agency
ENB	expected net benefit
EQ-5D	EuroQol EQ-5D questionnaire (for utilities)
ERG	evidence review group
EVI	expected value of information
EVPI	expected value of perfect information
EVPPI	expected value of partial perfect information
EVSI	expected value of sample information
FDA	Food and Drug Administration
GLM	general (or generalised) linear model
HAQ	health assessment questionnaire
HbA1C	glycosylated haemoglobin A
HDL	high-density lipoprotein
HRQoL	health-related quality of life
HTA	health technology assessment
ICC	intra-class correlation coefficient
ICER	incremental cost-effectiveness ratio
INB	incremental net benefit
INMB	incremental net monetary benefit
IPW	inverse probability weight
IQWiG	Institute for Quality and Efficiency (Germany)

ITC	indirect treatment comparison
ITT	intent to treat
IV	intravenous
LC-14	Lung Cancer Symptom Module of EORTC-QLQ-C30
MANOVA	multivariate analysis of variance
MDT	multidisciplinary team
MLM	multilevel model
MTC	mixed treatment comparison
NCLA	National Lung Cancer Audit
NHS	National Health Service
NICE	National Institute for Health and Care Excellence
NIHR	National Institute for Health Research
OLS	ordinary least squares
OS	overall survival
PFS	progression-free survival
PP	per protocol
PSA	probabilistic sensitivity analysis
QALYs	quality-adjusted life years
QLQ-C30	EORTC-QLQ-C30 quality of life in cancer questionnaire
QoL	quality of life
Q-TWiST	quality of time spent without symptoms of disease and toxicity
RAP/PAP	reimbursement/payer analysis plan
RCT	randomised controlled trial
REC	research ethics committee
SABA/LABA	short-/long-acting beta agonist
SAP	statistical analysis plan
SCET-VA V1.1	sample size and value of information software for cost-effectiveness analyses
SMC	Scottish Medicines Consortium
TEAE	treatment-emergent adverse events
TTH	time to healing
TTO	time trade-off
TTP	time to progression
TWiST	time without symptoms and toxicity
VOI	value of information
WTP	willingness to pay

1

Introduction to Economic Evaluation

1.1 Health Economics, Pharmacoeconomics and Economic Evaluation

Health expenditure in the United States passed $3,000,000,000,000 ($3 trillion) in 2014 and now accounts for just over 15% of US gross domestic product (GDP), according to *Forbes Magazine* (2014). Indeed, it is forecast to account for $3.6 trillion (about one-fifth of US economic activity) by the end of 2014 and may reach 33% of GDP by the year 2050. In the United Kingdom, health expenditure comprises about 17% of government spending. In every developed economy, healthcare is a major component of spending, investment and employment. Indeed, a healthy nation is crucial to the overall economic well-being of a country and its citizens.

Concerns have been raised as to whether healthcare should be considered an economic good (Morris, 2006). Some think these concerns are unfounded because, ultimately, healthcare resource is finite and scarce (Morris, 2006; Santerre & Neun, 2009). There is a limited supply of doctors, nurses and healthcare staff, so it is unlikely that any healthcare system in the world can achieve sufficient spending on the health of its citizens to meet all their healthcare desires. Some controversially argue that by not treating criminals, smokers and alcoholics, we can achieve healthcare for most individuals, whereas others argue that healthcare is a basic human right and to treat it as an 'economic good', like any other sort of consumer good or service, is not appropriate (Morris, 2006). We leave these questions to economic philosophers and policymakers; they involve complicated judgements which need not concern us. This book is only concerned with the quantitative methods used to determine the comparative value offered by new or existing treatments, and not with how they should be rationed.

Constraints on healthcare resource, particularly during times of economic turmoil and instability, result in governments taking a hard look at all public expenditure, including the medicines budget. Policymakers have a limited budget with which to decide how healthcare resource is provided for their citizens. In practice, 'payers' (i.e. governments or health insurance providers who contribute towards the cost of healthcare provision for citizens or customers) are likely to be more selective and *choose* with greater care from the healthcare options (i.e. new treatments) available to patients due to budget constraints. Just as most individuals cannot have all the things they want because of scarce resources (i.e. money, time), healthcare systems experience similar constraints when trying to meet the demands of their consumers (patients).

Economics is the science of scarcity and choice. *Health economics* is related to the supply and demand for healthcare. Although *health* itself is not an economic good, healthcare *resource* is. Healthcare resource refers to items such as hospital beds, treatments, drugs, surgeries, general practitioner (GP) time and so on. These resources usually have costs associated with them. For example, a visit to the doctor might be valued at £100 per hour. In this sense, health economics might be concerned with more than just medicines. Government institutions might make decisions at the national level on how the total healthcare demand and supply can be met for a given budget constraint.

However, what is usually of concern in the pharmaceutical context for comparing new treatments is *pharmacoeconomics*. Pharmacoeconomics uses certain principles of health economics for making policy decisions on the supply and demand for medicines – particularly in clinical trials. The methods (analysis techniques) used in pharmacoeconomics involve the 'description and analysis of the costs of drug therapy to healthcare systems and society. It identifies, measures and compares the costs and consequences of pharmaceutical products and services' (Rascati, 2013). The process of using these methods is called an *economic evaluation*. Economic evaluation applies mathematical and statistical methods to compare the costs and consequences of alternative options (Drummond et al., 2005).

For over a decade, economic evaluation has become an increasingly important component of clinical trials. The number of Phase III clinical trials with a health economic component has also increased substantially. In particular, submissions to reimbursement authorities – that is, government-led subgroups which assess the evidence for the 'value' that a new treatment offers, such as the Scottish Medicines Consortium (SMC) and the National Institute for Clinical Health Excellence (NICE) in the United Kingdom – have also increased by more than 45% over the same period.

In this book, we consider economic evaluation in the context of pharmacoeconomics, that is, evaluating the costs and benefits of treatments from clinical trials in which data are collected prospectively. The evaluation or assessment of the costs relative to the effects of specific treatments is made by special committees or government bodies, depending on the country. For example, in the United Kingdom (England and Wales), this assessment is made by NICE. Economic evaluation uses various techniques to assess the value of pharmaceutical interventions and health strategies. In short, most of pharmacoeconomics can be encapsulated by the single term *cost-effectiveness*.

A clinical trial with an economic evaluation is likely to involve individuals from several disciplines, including marketing, clinical research, biostatistics, health economics and epidemiology, who form part of a project team to develop a plan by which the new treatment can demonstrate value, not only for patients but also for the taxpayer, so that successful reimbursement can be achieved. Reimbursement, in simple terms, means the price the pharmaceutical company would like to obtain from the decision maker (payer) for the new drug treatment it has produced. For example, the pharmaceutical company might want £120 per tablet, but the payer might want to pay just £95 per tablet, based on the assessment of evidence of value presented by the company.

Figure 1.1 shows how the functions act together in an attempt to put together a value proposition. The value proposition is a documented strategy which outlines the plan for how the new treatment will show cost-effectiveness and has implications for how clinical trial data are collected and consequently analysed in the clinical trial.

Figure 1.1 also shows how a multidisciplinary team (MDT) might work together when designing a Phase III trial with an economic evaluation component. Statistics and clinical research groups would collaborate and design a well-powered study for the key clinical end points. The market access team – supported by the health economics and marketing groups – provides

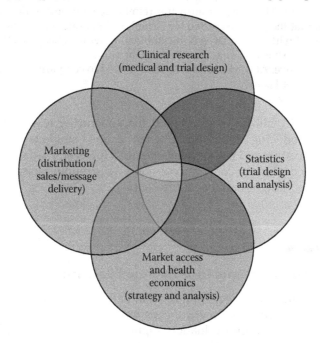

FIGURE 1.1
Description of a multidisciplinary team for designing a clinical trial with a health economic component.

expert input to identify characteristics of the new drug that are likely to demonstrate the 'value' argument and consequently increase market share (e.g. by suggesting secondary end points or analyses which are both clinically and commercially important if and when compared with a comparator treatment). For example, the market access and clinical teams might discuss what is needed to demonstrate cost-effectiveness: they can offer useful advice on the choice of a comparator likely to be acceptable to a given country. Statisticians and health economists might team up to discuss and determine pre-specified analyses (e.g. to agree on modelling techniques) which ensure the best chances of successful early reimbursement in the event that the clinical trial meets its objectives (generally, that the new treatment is better than the standard treatment).

Example 1.1: The Importance of Early Health Economic Input

Company Z has recently marketed an opioid (Drug X) hailed as an excellent innovation. The drug is used to treat non-malignant pain. One of the side effects of opioid treatment is constipation, often termed *opioid-induced constipation* (OIC). The advantage of Drug X is that the severity of constipation is less than with standard opioids. Drug X provides equivalent analgesic relief but less severe constipation than Drug S (standard). Despite this benefit, company Z was unable to secure a premium price for its drug. This was because the reimbursement authorities determined that the price requested for Drug X did not offer sufficient value compared with taking Drug S in combination with a cheap laxative (to treat OIC). That is, Drug X is not of great value versus Drug S plus a cheap laxative. At the time, no trial data existed to compare Drug X versus Drug S + laxative.

In Example 1.1, clinical research and manufacturing teams were clearly focused on producing a clinically excellent drug (a single tablet or capsule with both a laxative and a pain relief benefit). However, the value argument was overlooked. Had the company considered the reimbursement argument earlier, the story might have been different.

1.2 Important Concepts in Economic Evaluation

The different methods of economic evaluation for future decision- and policymaking are based on theoretical economic concepts, which will now be introduced. An economic evaluation might be carried out by analysts without appreciating its economic rationale. In the same way as a researcher/analyst needs a basic understanding of pharmacokinetics when designing Phase I clinical trials (e.g. the meaning of parameters such as C_{max} and area under the curve [AUC]), the same is true of the researcher or analyst who performs an economic evaluation: it is therefore useful to have an appreciation of some important economic concepts and terms when performing an economic evaluation.

1.2.1 Value

When it comes to healthcare, the 'average' person has no idea of the price for surgery or a treatment plan, especially when healthcare is considered to be 'free of charge', as in the National Health Service (NHS) in the United Kingdom. We might all be aware of the price of diamonds or footballs, and most of us can value them. Many people have some idea of the price of diamonds, and would pay a premium price for such goods. But it is likely that in a drought someone might exchange their diamonds for a cup of water (due to its scarcity).

Health products (or services) are not items that we can just buy 'off the shelf'. Therefore, it is harder to value and subsequently put a price on them. This is true whether the health item is a new treatment or something as complex as surgery. Some economists suggested that the problem of value could be determined by how much one was prepared to pay for a certain good, assuming certain market conditions were satisfied. Complicated economic theory was then formulated to explain how value could be determined by the demand for and the supply of a good (see Morris [2006] and Santerre and Neun [2009] for a useful introduction). The allocation of goods was determined simply by how much one was prepared to pay for that item (the price). In the context of health, the buyer is not necessarily the same as the consumer. The buyer of the health product is likely to be a government institution or an entity that is responsible for healthcare provision. The consumer is the patient. In other words, the allocation of healthcare goods is based on the price that governments (taxpayers) are prepared to pay. This is not necessarily true in all countries; it depends on the healthcare system in a particular country.

Valuing health, or for that matter a product, only according to how much people are prepared to pay does not take into account the impact on the wider society – or the *welfare* of everyone. For example, richer people are less likely than the poor to suffer from price increases. This was not considered equitable; therefore, in order to address this, *welfare economics* was developed.

One basic idea behind welfare economics was to provide a solution to situations in which, for example, the health resource used to treat *one* individual with Drug A could involve the same resource (money) as treating *three* individuals using Drug B. Society might judge the value of Treatment A to be insufficient compared with Treatment B. Welfare economics attempted to determine whether there were situations in which one could determine value based on the price of goods (e.g. price of Treatment A or B), but at the same time determine whether there exists an optimal distribution (of Treatments A and B) from a societal perspective. This optimum was manifested in an idea called *Pareto optimality*.

One key aspect of Paretian economics is that value can be based on the pricing mechanism (demand and supply) as long as, in situations where Treatment A made *one* patient better off, it did not make any other patient *worse off* (e.g. due to patients being denied Treatment B). That is, any further reallocation of Treatments A and B, although leading to one patient being better off, should not result in more than one other patient being worse off. Clearly, if the cost of A is £12,000, and that is all the resource available, then some other patients would indeed be worse off if the cost of Treatment B was £4,000 (because £12,000 is used for Patient 1, but nothing for Patient 2). If one individual is better off and no one else is worse off, this is called Pareto optimality. Of course, if the rich get richer and the poor do not get any poorer, this is also Pareto optimality, but it is not an ideal situation.

A further development in economics, called *extra welfare* economics, sought to address value in the health sector. This idea suggested that the economic evaluation methods which are used in this book are not really based on true welfare principles for various reasons we need not discuss here. The debate as to whether the current methods of economic evaluation are based on welfare economics can be found in excellent works such as Culyer et al. (2011) and Aki et al. (1998, 2001). What is important for our purposes is to appreciate that the economic evaluation techniques encountered in this book are tools for decision-making to determine which treatments or health technologies offer the greater value in terms of the price one is prepared to pay, termed the *willingness to pay* (WTP) or the *cost-effectiveness threshold*.

1.2.2 Allocative Efficiency

Allocative efficiency is where decision makers use the evidence (results) from an economic evaluation to determine the optimum set of allocations of treatments for a given medicines budget. For example, if only £100,000 were available for providing two treatment options, Drug A and Drug B, the question might be how best to spend the £100,000 on the treatments. If the price of Drug A is £2000 per year and the price of Drug B is £5000 per year, one could have 50 units of A or 20 units of B. The other option is to have *some* of A and *some* of B. Exactly what mixture of A and B to spend the £100,000 on is for the decision maker to choose. The *comparative* value that A and B offer will influence this decision.

The tools used in economic evaluation to influence this decision are described in Chapter 2. One such technique is *cost–utility analysis* (CUA), which seeks to address the question of allocative efficiency within the health sector. The idea is that patients will have access to 'value for money' treatments through an efficient allocation of various medicines subject to a budget constraint.

There might be a budget constraint (the amount of money available to spend) of £100 million to spend on cholesterol-lowering drugs for patients with high cholesterol, but since the drug might not work on every patient with high cholesterol, it might be more relevant to offer the resource to a specific group of patients (e.g. with specific demographic characteristics). For example, patients with more severe heart disease are more likely to benefit from a lipid-lowering drug than patients with a less severe condition. This might involve defining, for example, subgroup analysis in a clinical trial to identify where the new treatment might offer the greatest value.

1.2.3 Technical Efficiency

Technical efficiency is when the minimum amount of resources (e.g. lowest dose or shortest duration of dosing) is used to elicit a given response: for example, the lowest dose that achieves a 20% reduction in lipid levels, or the least costly way of resolving a peptic ulcer. *Cost-effectiveness analysis* (CEA) is one such tool used to address the issue of technical efficiency. The term CEA can be confusing, because CEA is also a general term used to describe how we determine efficiency. In this book, when the term *cost-effectiveness analysis* is used in its general sense, it refers to any one of the main tools used to determine efficiency, of which the most important are CEA, CUA, cost-minimisation analysis (CMA) and cost–benefit analysis (CBA), discussed later (Chapter 2). For this reason, the term *economic evaluation* is used instead of CEA whenever possible in order to distinguish between CEA as a specific tool and CEA as a general term for assessing the efficiency of healthcare interventions.

1.2.4 Reimbursement

In the context of pharmaceuticals, once a drug has been approved for licensing by regulatory bodies such as the Food and Drug Administration (FDA) or the European Medicines Agency (EMEA), the pharmaceutical company will seek a price for its newly licensed drug.

The price could be on a per tablet basis or for a supply of 28 days, such as £50 per tablet, or £1400 for 28 days. The price of £50 is what the payer has agreed to pay for each tablet. The price set is usually agreed between the payer (e.g. the Department of Health in the United Kingdom) and the pharmaceutical company. For example, the price per tablet of Lenolidamide (in 2011) was agreed at £249.60 (BNF 2011) for a single 25 mg tablet. This price is recorded in a publication called the *British National Formulary* (BNF). Reimbursement will be discussed again in Chapter 11.

The price is typically calculated to cover the research and development (R&D) costs of the pharmaceutical company as a minimum and to make a profit to sustain future R&D activities. From the payer's perspective, the lower the price, the better. However, the price should not be set too low so that innovation is discouraged. A premium price is usually a price higher than the current market price. The premium price may be awarded if the drug demonstrates value for money through economic evaluation techniques. Reimbursement is effectively the price reimbursed to the pharmaceutical company per prescription use of the drug. It is for this reason that when a health economic evaluation is undertaken, the payer perspective is considered carefully. That is, who is the economic evaluation for? Is it for a health insurance company (in the United States and some other countries, there is no equivalent of the NHS) or for the local (as in Spain, where local provinces can influence decisions) or national government?

1.2.5 Opportunity Cost

A very important concept in health economics, and economics in general, is *opportunity cost*. If the cost of 1 month's supply of an anti-cancer drug is £5000 for one patient, how else could the £5000 have been used instead of treating the patient? £5000 might also be the cost of 400 free school dinners, or of treating five patients with a similar drug (at £1000 per patient) but with a slightly smaller clinical effect (without harm). The cost of school dinners for 400 children, or five patients treated with the cheaper drug, is the opportunity cost. If all £5000 for the drug is given up for 400 free school meals, then 400 free school meals is the opportunity cost of the cancer drug.

Economic evaluation is often set against the background of an opportunity cost when comparing treatments. In the context of medicines, for a fixed budget of £100 million, for example, the payer may have the difficult decision of allocating all £100 million to Drug A for 10,000 patients, which might improve survival by 1 year (cost of £10,000 per year per patient). The opportunity cost might be spending £100 million on Treatment B for 20,000 patients, which might improve survival by 6 months (£5,000 per year per patient). In practice, a combination of Treatment A and B may be optimised.

1.2.6 Discounting

Most people prefer to receive benefits sooner rather than later, and pay costs later rather than sooner. For example, people exercise to keep fit so that they may have better *future* quality of life (QoL). Others prefer to smoke, and might prefer to enjoy smoking *now* and give less importance to their future health. These are examples of time preference. It is the rate of time preference that needs to be modified and termed *discounting*. As someone becomes older, for example, their time preference may change.

A common phrase is 'a pound today is worth more than a pound tomorrow'. This is because inflation plays a part in making a pound worth less tomorrow than today. In the same way, the costs of treatment in 5 years' time may be valued differently from the costs today. For example, the future costs of treatment for a single patient who experiences disease progression in each of years 1, 2 and 3 are expected to be £3,000, £5,000 and £7,000 (total £15,000 over 3 years). After discounting at 3%, the future costs are valued as

Year 1 $£3000 \times 1/(1+0.03)^1 = £2,912.62$ after 1 year

Year 2 $£5000 \times 1/(1+0.03)^2 = £4,712.98$ after 2 years

Year 3 $£7000 \times 1/(1+0.03)^3 = £6,405.99$ after 3 years

The total future costs of £15,000 for this patient after discounting are £14,031.59. In this example, only costs are discounted. In practice, health benefits are also discounted. The debate about whether or not we should discount costs and not benefits is discussed elsewhere (e.g. Drummond, 2005) and not considered further here. The current practice, however, is to discount both future health benefits and costs.

In health economic evaluation, future costs and benefits are discounted so that their value can be judged in present terms. This is achieved by applying an annual discount rate, for example, 3%. The discount rate in the United Kingdom is based on the Treasury Department's so-called Green Book. The consequence of the discount rate is that less weight is given to later costs than to present ones. We use the discounted values of future costs and benefits for further calculations to determine cost-effectiveness.

When no future costs and benefits are expected in a clinical trial, discounting is not usually applied. For example, at the end of a 3-year clinical trial (conducted between 2008 and 2011), the estimates of costs might be taken from year 2011 list prices and assumed to be the same for years 2008, 2009 and 2010 (taking inflation into account). If a GP visit for a particular patient in 2011 costs £150, then we may assume that it is also £150 in the other years of the trial for other patients, ignoring inflation (Gray et al., 2011). Another possibility is to adjust costs which occurred between 2008 and 2010 to the current (2011) year equivalent to using a consumer price index (which takes inflation into account).

When discounting occurs for future benefits and costs at the patient level, there can be added complexities. For example, in oncology trials, the follow-up period is often limited, and estimates of future benefits (e.g. overall life years expected, measured in quality-adjusted life years [QALYs]) are sometimes modelled using aggregate (total) values, whereas future costs and effects occur at the patient level. Example 1.2 shows how discounting can be carried out in practice with patient-level data.

Example 1.2: Discounting Costs and Benefits

Table 1.1 shows an example of discounting in three patients for Treatment A (assume a two-group trial, with Treatments A and B). All three patients were followed up in the trial for 3 years. The total costs during the trial

TABLE 1.1

Discounting Future Costs and Benefits with Patient-Level Data for a Single Treatment Group

| | During the Clinical Trial | | | | | Future Costs and Benefits | | | | | | Overall | |
| | Costs (£) | | | | Benefits[a] | Costs (£) | | | | Benefits | | | |
Patient Years	Y_0-Y_1	Y_1-Y_2	Y_2-Y_3	Total	Total	Y_3-Y_4	Y_4-Y_5	Total	Discounted	Y_3-Y_5	Discounted	Costs	QALY
1	500	1000	3000	4,500	2.5	2000	1000	3000	2942	1.1	1.06	7,442	3.56
2	500	1000	2000	3,500	2.8	1000	1000	2000	1942	1.3	1.22	5,442	4.02
3	400	3000	1000	4,400	1.6	800	900	1700	1648	1.7	1.55	6,048	3.15
Total	1400	5000	6000	12,400	6.9	3800	2900	6700	6532	4.1	3.83	18,932	10.73

[a] Benefits measured as QALYs.

were £4,500, £3,500 and £4,400 for each patient over 3 years, with a total undiscounted cost of £12,400. These costs could include drug costs, cost of hospitalisation and so on. The total present (£12,400) and future (£6,700) undiscounted costs are £19,100 over the 3 years of the clinical trial (Table 1.1). The total discounted future costs over 2 years are £6532. For Patient 1, this is £2000 + 1000 × $(1.03)^2$ = £2942. Therefore, for Patient 1, the total costs are £500 + £1000 + £3000 + £2000 + £942 = £2942 after 3% discounting. Since the future period of time is only 2 years, the impact of discounting in the second year has been minimal.

The benefits are measured in QALYs, which combine QoL and quantity of life (Chapter 5) into a single metric of quality-adjusted survival. For each of the three patients, follow-up was over 3 years during the trial. At the end of the trial, the QALYs were computed as 2.5, 2.8 and 1.6 for the three patients, respectively. However, patients were expected to survive beyond the trial follow-up time. Therefore, these benefits (survival and QoL) over a further 2 years were considered to be future benefits which should also be subject to discounting. In particular, Patient 1 was expected to yield a future QALY of 1.1 over 2 further years, and after discounting, this would be 1.06. The total undiscounted QALYs over the 2 years across all three patients were, therefore, 4.1 and 3.83 after 3% discounting.

In economic evaluation, the discounted values are those used for future statistical modelling and analysis. Hence, in Table 1.1, the values used in a patient-level analysis would be £7442 and 3.56, £5442 and 4.02 and £6048 and 3.15 for Patients 1, 2 and 3 for costs and effects, respectively.

Future benefits are often estimated through modelling techniques (Chapters 6 and 10). We shall return to these types of calculations later. For now, it should be clear how the discounting is applied to estimate future costs and QALYs.

1.3 Health Economic Evaluation and Drug Development

Figure 1.2 displays the traditional drug development process and approaches to reimbursement and market access. The dashed outline shows how the role of pharmacoeconomics has reshaped this process. In the past, less effort would be planned for demonstrating the value of a new treatment. The traditional route was to perform Phase I to Phase III trials, obtain a market authorisation licence and then agree with each country (separately) a price for the new treatment. Evidence of value would not have been formally requested. Only efficacy concerns were considered important at the time of pricing (not relative efficacy or costs). Therefore, one role of market access would be to agree with the country's government body a price at which they would buy

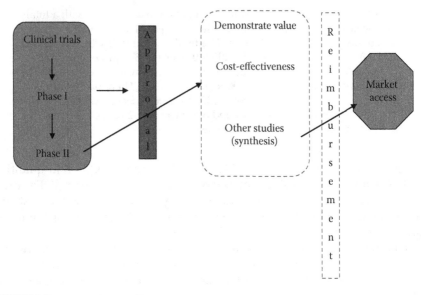

FIGURE 1.2
Drug development and reimbursement.

the new drug – agreed often relatively shortly after the drug was approved. The evidence for informing a pricing decision was based primarily on using the Phase III clinical trial data submitted for market authorisation.

In Germany, for example, the concept of 'free pricing' allowed innovator companies to exert greater control over price and set a price with considerable flexibility. However, the AMNOG law in 2011 (Neuordnung des Arzneimittelmarktes [Restructuring of the Pharmaceutical Market]) effectively restricted free pricing to a 1-year period only, and the pharmaceutical company is required to perform an assessment of value for money of the new treatment within the first year. German payers no longer find it acceptable to pay for expensive drugs which are seen to offer little value for money. In particular, oncology drugs are likely to feel the impact of the Institut für Qualität und Wirtschaftlichkeit im Gesundheitswesen (The Institute for Quality and Efficiency, also known as IQWiG) decisions more sharply, because some of these drugs are particularly expensive and reviewed judiciously from the perspective of demonstrating value. Previously, approaches to market access were not influenced by the concept of value for money. There was less need to formalise the health economic argument and package the data in a way which demonstrated the uncertainties of value for money.

The new drug development paradigm requires formal evidence of the cost–benefit/value relationship. An MDT is set up which considers as early as possible what is needed from the clinical trials data to form a value-added argument. Although the analysis of data for economic evaluation occurs after the Phase III trial results are finalised, the design and planning for both efficacy and showing value are considered well before then. The MDT bridges the working relationship between individuals from clinical research, biostatistics health economics and other disciplines to formalise the evidence from clinical trials to obtain market access by demonstrating value.

A strategic partnership between all researchers involved in drug development can result in fruitful results when attempting to demonstrate or plan the 'value-added' aspects of new drugs, often needed for price reimbursement through cost-effectiveness arguments.

Demonstrating efficacy in a randomised controlled trial (RCT) is not necessarily the only criterion for reimbursement and successful market access (Figure 1.2). Greater synthesis of existing data (Sculpher, 2006) is required so that all available information is used before a decision regarding the value of a new treatment can be made. The implication of Figure 1.2 is that if the innovator drug cannot demonstrate value for money to the payer, the price desired may not be achieved. This does not mean that a drug is not efficacious without such synthesis, but the decision-making process should use *all* available information to minimise uncertainty.

However, if all 'other' information is incorporated, it is not clear how this should be weighted or, more importantly, whether it creates additional uncertainty, especially when the 'other information' is of poor quality and very heterogeneous. In this case, it is possible that the price of the new drug may be set equal to the price of a cheap generic and much lower than the 'hoped-for' price. This naturally impacts reinvestment, shareholder returns and the profit margins of pharmaceutical companies. It is therefore important that the planning of value in a clinical development programme is carefully and diligently executed. Early estimates of relative costs to effects expressed in terms of incremental cost-effectiveness ratios (ICER), discussed in Chapter 2, might be useful to understand where the new drug lies in terms of market positioning. Careful and efficient clinical trial designs which do not compromise either the clinical or the economic objectives should be considered (Chapter 3). For example, requesting early trial data to get an early idea of the ICER to develop a reimbursement strategy may result in unblinding, which compromises the primary end point of the trial.

The relationship between the drug development programme over time, market authorisation (getting a licence), market access (as measured by sales volume) and the importance of reimbursement is shown in Figure 1.3. After licence approval, there is a period of time between approval and a decision for reimbursement. During the period between market authorisation and reimbursement, for most drugs, sales are still somewhat flat. This is because either the reimbursement authorities are undecided as to whether the new treatment offers value or a decision is pending. However, when a decision has been made by reimbursement agencies that the new treatment does offer value for money, the sales are likely to increase (difference between K and W in Figure 1.3) because the new treatment has more value and would be recommended for use. The premium price agreed would also influence profits.

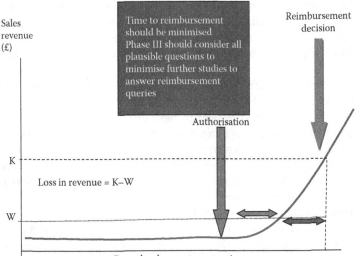

FIGURE 1.3
Relationship between drug licensing, market access and reimbursement (W: sales at some amount £W are flat, after which £K are achieved [after reimbursement]. Consequently, the loss in revenue is £K−£W).

The loss in revenue that occurs during the period between market approval and the reimbursement decision is not a trivial one. On one occasion with a sponsor, several analyses were deliberated during the clinical trial because statisticians were debating as to whether exploratory analyses which were unspecified in the statistical analysis plan or protocol could be performed. While this may be a concern for the regulatory (market authorisation) agency for reimbursement authorities, it was not a concern for a reimbursement agency (statistical significance issues such as inflation of Type I errors are not of primary importance in economic evaluation, and statistical inference can be irrelevant, as we will note later [Klaxton, 1999]). Consequently, the reimbursement decision was delayed, resulting in millions of euros of lost revenue.

Therefore, it may be prudent at the clinical development programme stage to consider pharmacoeconomics earlier to plan a dossier for the reimbursement. Even if the drug is not efficacious, the risk of losing substantial market access is easily offset against the resource needed to prepare a value argument.

1.4 Efficacy, Effectiveness and Efficiency

The concept of efficacy is well understood in the context of an RCT. Restricted inclusion criteria and very controlled monitoring of efficacy and safety outcomes give the RCT framework a 'gold standard' status in terms of internal validity. The possibility of bias is considered to be well accounted for (Pocock, 1983). However, the RCT provides less external validity, because any inferences are restricted to the population under investigation, and lacks generalisability (Sculpher, 2006).

Figure 1.4 shows the relationship between the drug development process and the activities related to an economic evaluation during each phase of drug development.

Although one key objective of pharmaceutical companies is to carry out scientifically robust trials, there are also obvious concerns about profitability and risk throughout the drug development process. For example, in the development of a transdermal patch for pain, during early drug development, concerns around manufacturing the patch and comparing it with competitor patches in terms of delivering dose loads and other practicalities (e.g. patch does not fall off) are important, in addition to demonstrating superior efficacy.

During the early and preclinical phases of drug development, the market landscape for a new treatment is assessed. By Phase II (and increasingly by Phase I), early evidence of efficacy will allow optimisation of early cost-effectiveness models developed at the earlier stages (preclinical and Phase I). Phases III and IV involve revising the cost-effectiveness models developed during Phase II as well as preparing the value arguments. Finally, at launch and post Phase III, the impact of the models on profitability and risk (P&R) as well as evidence required for local markets is determined and submitted.

What is clear from Figure 1.4 is that health economic–related activities do not start at Phase III, but are considered throughout the drug development process. Although the objective of a Phase III trial is efficacy, an early measure of cost-effectiveness could be useful to plan pricing and consider what else is needed for the value argument.

Effectiveness, on the other hand, is demonstrated by testing a new treatment in less restrictive (real-world) conditions. Another objective of a clinical trial in which cost-effectiveness is a key component should be to measure some aspect determining the clinical effects of the new treatment in a real-world setting (if feasible). Measuring efficacy in patients with co-morbidities, longer-term follow-up, less controlled dosing and poorer patient compliance may not have been considered in the Phase III RCT. The target population of patients may be broadly similar to those in the clinical trial, but additional questions such as 'How well does the drug work in real practice?' and 'How well does the new treatment perform over a longer duration?' need to be answered. The RCT offers a good opportunity to collect these data if carefully designed. For example, a clinical trial follow-up period might be suggested as 12 months after the first dose of study treatment; however, a follow-up of 24 months might offer an opportunity to collect data to gauge a sense of how the new treatment is working in a real-world setting when double blinding and some of the other restrictive conditions are relaxed.

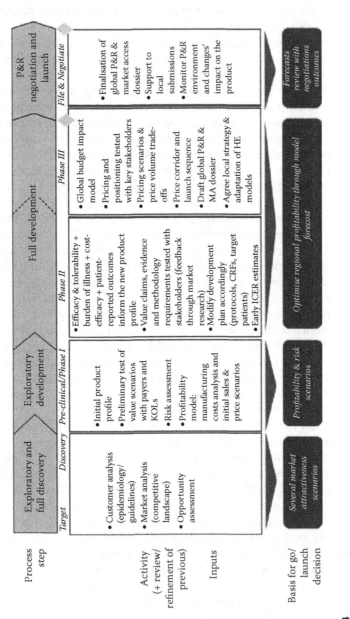

FIGURE 1.4
Relationship between the drug development process and health economic–related activities.

In addition, a measure of compliance over a longer period would provide valuable information on the use of the new treatment in practice, especially maintenance therapy, which in cancer trials can be particularly expensive, causing the ICER to be very large.

In an RCT, compliance is often closely monitored, and high protocol-driven compliance rates may be artificial. True compliance may be as low as 60%, well below some commonly stated compliance rates of 80% (often suggested for per protocol analyses). Although a per protocol population (a population of patients who are deemed to have complied with the protocol as far as possible) has an impact on the efficacy end points, the effect of protocol violators on costs is not often considered (Ordaz, 2013; Briggs, 2003; Noble, 2012). The intention to treat (ITT) population, which usually includes all patients randomised (as a minimum), is not always useful for assessing effectiveness, because the populations are still restricted by inclusion/exclusion criteria and, in some cases, the ITT is simply defined as 'all patients randomised' regardless of whether they have taken the treatment. In practice, clinical trials designed for cost-effectiveness will attempt to capture resource use and associated costs (Chapter 4) associated with each treatment group.

Table 1.2 shows the relationship between study objectives and key features for particular types of studies. The 'gold standard' to address confirmatory efficacy is the RCT with primary efficacy and safety outcomes; the time frame can be long term or short term although longer trials (e.g. cancer, cardiovascular and mortality end point) can become expensive to run. A pharmacoeconomic study, on the other hand, is best suited to evaluate efficiency, and a combination of evidence from RCTs and observational studies can be used. Outcomes such as resource use (costs), QoL and compliance, for example, are collected in such studies to assess efficiency. In practice, there may be a hybrid-type approach that optimises the potential to do as much as possible in a single trial.

TABLE 1.2

Relationship between Types of Study and Design Features

	Study Type		
	Clinical Trial	**Outcomes Research**	**Pharmacoeconomic**
Objective	Efficacy	Effectiveness	Efficiency
Design	RCT	Observational and RCT	RCT, observational and various others
Measures	Efficacy and safety	Patient-based outcomes	Resource use and outcomes
Time frame	Short term or long term	Long term	Long term
Based on	Ideal clinical practice (restricted population)	Normal clinical practice (wider population)	Normal clinical practice

1.5 When Is a Pharmacoeconomic Hypothesis Possible?

Table 1.3 shows the relationship between the potential clinical advantage envisaged in a clinical trial and how this can translate into a potential pharmacoeconomic hypothesis to demonstrate value. It is unlikely that a pharmacoeconomic hypothesis can be postulated unless some form of clinical advantage is plausible. In some clinical trials, no clinical advantage is possible, such as 'equivalence' trials, in which treatment benefits are considered to be similar to, or not worse than, those of a standard treatment. Bioequivalence trials are also equivalence trials, and even though there might be a change in the mode of administration (e.g. where absolute bioavailability is required for intravenous vs. oral dosing), this type of trial (with very few subjects) is unsuitable for any economic evaluation, because healthy volunteers are used and clinical benefit is not assessed.

1.5.1 Hypotheses

1.5.1.1 Superiority

A superiority trial is one in which the new treatment is clearly superior to the standard. The mean treatment difference, Δ_{A-S}, where A is the new treatment and S is the standard treatment (the symbol Δ_{A-S} represents a numerical value for the mean difference between treatments A and S), is a value such that the lower (or upper limit) of its 95% confidence interval (CI) excludes the value 0. When this happens, a new treatment is said to be 'superior' to the comparator (a 95% CI for the mean difference of [1.2–6.2] is statistically significant because the value 0 is not in this interval).

The value of Δ_{A-S} should be large enough to postulate a cost-effectiveness hypothesis. The value argument may depend on observed differences in *mean* costs between Treatments A and B relative to the mean difference in

TABLE 1.3

Relationship between Clinical Objective and Plausible Pharmacoeconomic Hypotheses

Clinical Advantage	Possible Pharmacoeconomic Hypotheses
Superior efficacy	Saves life years
	Averts disease
	Improved QoL/QALY gain
Better side-effect profile	Improved QoL
Change in half-life	More convenient administration
	Improved compliance
	Improved QoL/QALY gain
Improved delivery	Improved compliance
	Improved QoL/QALY gain

costs. For example, a mean difference of 0.3 mmHg in a large trial (n = 2000 patients) might be statistically significant with a 95% CI of (0.1–0.5). The value of Δ_{A-S} is only 0.3 mmHg, but whether this difference is large enough to demonstrate cost-effectiveness is a separate question. This is an example of a large trial with a small treatment benefit which is statistically significant, but may not necessarily yield a cost-effective clinical benefit. On the other hand, even if Δ_{A-S} is large, but the costs associated with this benefit are also high, then a cost-effectiveness argument may still not exist, because the difference in costs may be too high relative to clinical benefit.

Example 1.3: Statistical Significance versus Economic Relevance

In the example described in Section 1.5.1.1, a difference of 0.3 mmHg relative to *differences* in costs of £10,000 (i.e. Drug A is £10,000 more expensive than S), the ratio of this benefit is £10,000/0.3 > £33,000 (i.e. the new treatment costs £33,000 on average, for every one unit reduction in millimetres of mercury, for each patient). On the other hand, if the mean difference in effect is twice as large (i.e. 0.6 mmHg in favour of Treatment A) but the cost is £25,000 higher on Treatment A than B, this ratio is even worse for Treatment A (£25,000/0.6 > £41,000).

1.5.1.2 Non-Inferior Equivalence

A non-inferiority trial is one in which the average treatment benefit of the new drug is not considered to be clinically 'worse' than an existing treatment. In this case, the new drug is considered 'equivalent and not worse' compared with the standard treatment, as long as the lower (or upper) limit of the 95% CI for any observed difference lies above (or below) some predefined value −K (or) +K (in Figure 1.5, these values are −4 and +3.4, respectively). If observed treatment effects are likely to be lower (higher) than the value of K, the new treatment can be considered non-inferior. The definition of 'worse' or 'non-inferior' is subjective, based on clinical opinion.

There are two particular cases, important for economic evaluation, which concern non-inferiority: (i) where $\Delta_{A-S} > 0$ and (ii) where $\Delta_{A-S} < 0$. In case (i),

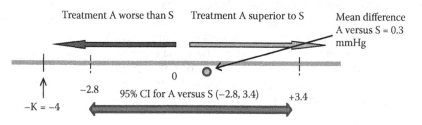

FIGURE 1.5
Example of non-inferiority result for case (i): average improvement of A vs. S = 0.3 mmHg, which is not enough to merit superiority, but is non-inferior to Treatment S because the lower 95% confidence limit of −2.8 is above the non-inferiority limit of −4.

the average treatment benefit is positive (A is better than S, on average), whereas in case (ii), the mean effect is worse for Treatment A. The value of Δ_{A-S} is not large enough to merit superiority in (i), but nevertheless there is some small clinical advantage (on average). Figure 1.5 illustrates case (i).

Example 1.4: Superiority and Noninferiority in the Context of Cost-Effectiveness

Consider a respiratory study in which the observed treatment difference (A vs. S) in blood pressure is 0.3 mmHg. The 95% CI for this difference is −2.8 to +3.4 mmHg (see Figure 1.5). For a conclusion of superiority, the 95% CI for the mean difference (Δ_A−S) would need to exclude the value 0 *and* show a positive treatment benefit. In this example, the magnitude of the benefit, on average, is 0.3 mmHg, but the condition for superiority is not satisfied (i.e. the value 0 is in the 95% CI). The lower limit of the 95% CI is −2.8 mmHg. This value is above the threshold limit of K = −4 mmHg (the condition for non-inferiority has been met). Had the lower 95% CI for the observed difference fallen below K = −4 (solid line in Figure 1.5), Treatment A would not have satisfied the non-inferiority conditions.

In Example 1.4, the average treatment effect showed some small benefit, despite not being superior. With such a small (potentially clinically irrelevant) benefit, it might still be plausible that the QoL enjoyed, on average, with Treatment A was better than with Treatment S. It is also important to note that from a purely clinical perspective, if a new treatment offers a small amount of benefit, then this additional benefit is considered to be *clinically irrelevant* because the objective is one of non-inferior equivalence.

From an economic evaluation perspective, however, the irrelevance of this small benefit may not always be negligible. Economic evaluation assessments are based on mean effects. The 95% CI for the mean difference is not considered an important metric for economic evaluation (statistical inference is irrelevant here; Claxton, 1999). Methods for handling uncertainty of point estimates are considered later under sensitivity analysis (Chapter 7). Economic evaluation considers the mean effects in its calculation of health benefit relative to costs. It is, therefore, possible that a new drug can be cost-effective, even if the objective is non-inferior equivalence. If there are other additional advantages, such as better safety and compliance, the economic argument for the new (but non-inferior) treatment becomes stronger. The value argument would not be based on efficacy alone, but on other considerations also. In this section, the point is made that a clinically irrelevant benefit which falls (just) short of superiority in a non-inferiority trial may still offer opportunities for showing economic benefit.

1.5.1.3 Non-Inferior Equivalence Where $\Delta_{A-S} < 0$ (Case II)

In this situation, the objective is again non-inferiority; however, the new treatment is not only shown to be non-inferior to the standard treatment, but also, on average, shows a worse effect (the difference A vs. S, Δ_{A-S}, is either 0 or negative [$\Delta_{A-S} < 0$]). From a clinical perspective, the fact that Treatment A is on average worse by a small margin is also considered clinically irrelevant. The degree of 'worseness' is not based on the mean difference, but on the 95% lower confidence limit. In this situation, as far as the new treatment is concerned, a clinical advantage is unlikely to exist. If there is a value argument, it is likely to be based on 'equivalent' treatment benefit and lower costs (or better safety).

Example 1.5: Equivalent Effects in the Context of Cost-Effectiveness

For the treatment of infection, a twice-a-day regimen is currently standard. A new once-a-day modified-release formulation is developed, which is a more convenient form of administration. The value argument might be based on showing that 'once a day versus twice a day' is likely to lead to better compliance and is cheaper. The manufacturer would seek a premium price as a result of this added value. In addition, the treatment effects might be similar, or perhaps even inferior (although this is unlikely), with the once-a-day regimen. Since the costs associated with the new (once-a-day) regimen are likely to be lower, the formulation with the lower cost is likely to be more efficient (efficacy assumed to be similar). An example of this situation might be a twice-a-day form of clarithromycin (an anti-infective drug) versus a modified (once-a-day) formulation.

Cost minimisation (Chapter 2) has in the past been one method to carry out an economic evaluation (Drummond, 2005) when treatments are equivalent or non-inferior. In the past, statistical testing for differences was carried out on costs and then separately on effects. If no differences were shown, equivalence was concluded. Subsequently, researchers suggested: 'just because there is no difference, it doesn't mean they are equivalent' (Senn, 1997; Briggs & O'Brien, 2001; Dakin & Wordsworth, 2011).

Using Example 1 from Briggs & O'Brien (2001), we consider an example in which a trial is not designed as a non-inferiority trial. This example is sufficient to demonstrate the points made earlier (Section 1.5.1.3).

Example 1.6: No Statistical Difference in Costs, but Important Cost-Effectiveness Conclusions

The mean survival time of a new drug for cancer (N) was 4.35 years (in economic evaluation the mean is the statistic of choice, whereas the median is reported in clinical trials) compared with 4.58 years with the standard treatment. The difference between mean effects was, therefore, 0.23 years, and for mean costs, the difference was $4911 (Treatment N was more expensive). There were no statistical differences between treatments for either the mean cost or the mean effect ($p > .05$). A conclusion of no difference was made. The ratio of costs to benefits was $4,911/0.23 = $21,352 per patient on average. This value had about a 70% chance of being cost-effective if the payer was prepared to pay $30,000.

TABLE 1.4

Summary of Hypotheses for the Primary End Point of a Trial

Hypothesis	Clinical Relevance	Average Difference (New vs. Standard) at End of Trial	Example of Possible Cost-Effectiveness Argument
Superiority	New treatment is better than standard	Improved with new $(\Delta_{N-S} > 0)$	Efficacy and possibly safety
Non-inferiority	New treatment is not worse than standard	Improved with new $(\Delta_{N-S} > 0)$	Possibly efficacy or safety, but mainly lower costs
Non-inferiority	New treatment is not worse than standard	New is worse $(\Delta_{N-S} < 0)$	Lower costs/better safety profile
Equivalence	New treatment is not worse or better than standard	New is neither better nor worse	Any of the above

Note: N: new; S: standard.

In sterling (£), the ratio would be sufficient to conclude a strong value argument at the NICE threshold of £30,000. That is, despite the trial being (incorrectly) considered as an equivalence trial, the new treatment (N) is cost-effective at a $30,000 threshold.

The point estimate of 0.23 years showed some benefit, even if the treatments were considered to be the same (i.e. not different).

Table 1.4 describes how the different clinical hypotheses can be translated into pharmacoeconomic hypotheses.

Exercises for Chapter 1

1. The primary end point from a clinical trial is the only outcome that is important for determining the value of the treatment. Discuss.

2. A trial that is designed for non-inferiority will never be able to demonstrate cost-effectiveness. Do you agree?

3. How would you decide on whether a new treatment was cost-effective in the following situations (assume two treatments being compared against each other)?

   a. The primary end point was very positive (i.e. a good outcome for the new treatment) and all secondary outcomes were also better for the new treatment.

   b. The primary end point was very positive (i.e. a good outcome) and all secondary outcomes were worse for the new treatment.

c. The primary end point was negative and all secondary outcomes were worse for the new treatment.

d. The primary end point was no different between treatments but all secondary outcomes were superior for the new treatment.

Is there a limitation in the way current economic evaluation is performed based on your answers to the above?

4. Why is market access important for a pharmaceutical company, and what does its success depend on?

2

Health Economic Evaluation Concepts

2.1 Incremental Cost-Effectiveness Ratio (ICER)

In economic evaluation, the main results are usually reported in one of two ways:

1. The incremental cost-effectiveness ratio (ICER)
2. The incremental net monetary benefit (INMB)

In this chapter, the relationship between ICER and INMB is explored. In Chapter 1, the ICER was informally introduced as costs relative to benefits. We now formally present the ICER in the context of the cost-effectiveness plane, which is how the results of an economic evaluation are often reported and interpreted.

The ICER is defined as

$$\frac{\text{Mean Costs}(A) - \text{Mean Costs}(S)}{\text{Mean Effect}(A) - \text{Mean Effect}(S)}$$

$$= \frac{\mu_A - \mu_S}{\varepsilon_A - \varepsilon_S} = \frac{\Delta\mu_{A-S}}{\Delta e_{A-S}} = \frac{\Delta_c}{\Delta_e} \tag{2.1}$$

where:

$\Delta_c = \Delta\mu_{A-S} = $ mean difference in costs between Treatments A and S
$\Delta_e = \Delta\varepsilon_{A-S} = $ mean difference in effects between Treatments A and S
$\mu_A = $ mean costs of Treatment A
$\mu_S = $ mean costs of Treatment S
$\varepsilon_A = $ mean effect of Treatment A
$\varepsilon_S = $ mean effect of Treatment S

The numerator in Equation 2.1, $\mu_A - \mu_S$, is called the *incremental cost*, and the denominator, $\varepsilon_A - \varepsilon_S$, is termed *incremental effectiveness*; A and S are two treatments (A is typically the new drug and S is the standard). It is this ratio quantity (Δ_c/Δ_e) and quantifying the uncertainty around it which lies at the heart of economic evaluation in clinical trials. This ratio is displayed on the cost-effectiveness plane shown in Figure 2.1.

In Figure 2.1, the x-axis represents incremental effectiveness or the mean difference in the effects between treatments A and S (Δ_e). For example, positive values of Δ_e ($\Delta_e > 0$) exist where the new drug is more effective. Effectiveness does not necessarily mean efficacy in Equation 2.1. It could be a measure of efficacy (e.g. survival time) combined with quality of life (QoL) to get a value in quality-adjusted life years (QALYs) (or life years gained or saved [see Chapter 5]). The value of Δ_e can also be measured in natural units (difference in blood pressure in millimetres of mercury [mmHg]) or as a proportion (or on a 0 to 1 scale). Negative values ($\Delta_e < 0$) indicate that the new treatment has lower effectiveness.

The y-axis in Figure 2.1 is the mean difference in costs (Δ_c), measured in some unit of currency (pounds sterling in this case). For example, if the new drug (Treatment A) costs £2000 more than the standard (Treatment S), the value of Δ_c is +£2000. If Δ_c is <0, the negative value means that Treatment A costs less than S, on average.

In Quadrants 2 and 4, the decision as to which treatment is more or less cost-effective is relatively easy. If the value of the ICER from Equation 2.1 lies in Quadrant 2, the new treatment is cheaper and more effective. This is the ideal scenario in which pharmaceutical companies would like to have their new drugs positioned. On the other hand, a less desirable scenario is where the new treatment is worse, but also more costly (Quadrant 4). The values of the ICER can, however, be altered by changing parameters such as the price

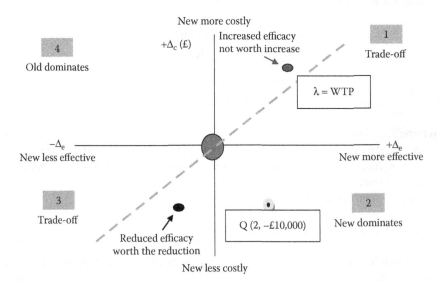

FIGURE 2.1
Cost-effectiveness plane.

of the new treatment. Reducing the price (or increasing efficacy, if possible) might be a strategy adopted so that the ICER can move into a different, more favourable quadrant, possibly at a reduced profit. A new treatment which has an ICER that falls into Quadrant 4 is unlikely to be considered as having a high chance of demonstrating value. Even if the price was changed, the fact that the new treatment has poorer efficacy still needs to be addressed.

Example 2.1: Interpreting the Cost-Effectiveness Plane

Referring to Figure 2.1, we note that in Quadrant 2, the point Q (2, −£10,000) shows that a new drug is more effective (an improved effect of 2 and cheaper by £10,000) but also cheaper (by £10,000), on average. Hence, the ICER is −£10,000/2 = −£5,000 per unit of effect (e.g. the unit could be QALYs). The new treatment is said to dominate the standard treatment. Most decision problems relating to the value of the ICER are concerned with Quadrants 1 and 3, and in particular Quadrant 1, where justification of value is often sought.

The line that passes through the origin (Figure 2.1), denoted λ, is called the willingness to pay (WTP) or cost-effectiveness threshold. This is the threshold ratio or amount in pounds (or other currency) that a payer would be prepared to pay for a new drug. Any ICER values calculated from data which are to the right of this line (e.g. in Quadrant 1) show that the new treatment is cost-effective. In this example (Figure 2.1), the incremental effect value (Δ_e) is 2 and the incremental cost value (Δ_c) is −£10,000, resulting in an ICER = −£5,000, shown as the point Q(2, −£10,000) in Quadrant 2. Had the new treatment showed poorer efficacy compared with the standard (e.g. a value of −2.8), the ICER would be −£10,000/−2.8 = £3,571. The ICER is now positive and has shifted from Quadrant 2 to Quadrant 3 (Figure 2 1).

Example 2.2: Changing the Cost-Effectiveness Threshold

Figure 2.2 shows two slopes: λ (dashed line) and λ^* (solid line) – the cost-effectiveness thresholds. The value of the cost-effectiveness threshold has changed from λ = £30,000 (dashed line) to λ^* = £12,000 (solid line). Initially, the new treatment (not the same as in Example 2.1) showed a treatment benefit of 2 units, but was more expensive (£28,000) – the point Z (2, £28,000) in Figure 2.2. The point Z initially lies below the line λ = £30,000, but as the value of λ = £30,000 changes to λ^* = £12,000, the point Z is above the new cost-effectiveness threshold. The observed slope, the ICER, is £28,000/2 = £14,000, which lies to the left and above the new cost-effectiveness threshold. At this new threshold, Treatment A is no longer considered cost-effective (because the point Z is above the line).

In general, in Quadrant 1, if $\Delta_c/\Delta_e \, \lambda$ and as long as $\Delta_e > 0$, the new treatment is cost-effective; values of $\Delta_e > 0$ suggest a benefit with the new treatment. In

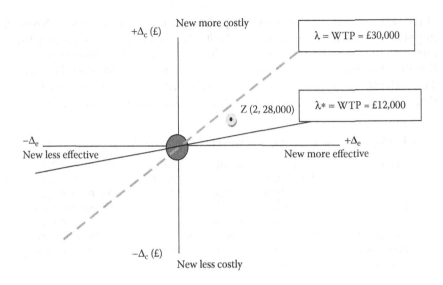

FIGURE 2.2
Cost-effectiveness plane: Changing WTP/CE threshold.

Quadrant 3, Δ_c/Δ_e is always ≥ 0 (for $\Delta_e \neq 0$), so the ratio is $< \lambda$, and the new treatment is considered cost-effective.

2.2 Incremental INMB

Previously, decisions relating to cost-effectiveness were made on the ICER and the uncertainty around it. The uncertainty was often expressed as a 95% confidence interval (CI) for the ratio $Y = \Delta\mu_{A-S}/\Delta\varepsilon_{A-S} = \Delta_c/\Delta_e$ (Briggs, 1997; Willan & Briggs, 2006). There were several problems associated with assessing the uncertainty of the ICER. First, Y is a ratio and has some awkward statistical properties which can result in difficult statistical inference. For example, the quantity Δ_e can be 0 or very close to 0 (for example, in equivalence trials), resulting in a very large or infinite ICER. In addition, if some ratios are negative, this makes the interpretation of Y trickier, especially when we need to compare more than two treatments, where often the ratios of several comparisons are ranked to identify dominant treatments (see Section 2.3 on dominance).

The main approaches to addressing how to provide a measure of uncertainty around the ICER are

1. Taylor's Expansion to estimate the variance of the ICER as a ratio of two quantities (involving a complicated equation)

2. Fieller's theorem, which can result in wider CIs for the ICER (discussed in Appendix 2A.1)
3. Bootstrapping, a resampling method to estimate the statistical properties of the ICER (see Example 2.5)
4. The INMB approach, which removes the difficulties of statistical inference

Another problem with the ICER, especially when using CIs, is its interpretation. Since the ICER can be negative, the 95% CI can have a negative lower limit with a positive upper limit. This is in itself not uncommon in clinical trials with efficacy end points. For example, a change from the baseline in a forced expiratory volume (FEV_1) analysis can yield a 95% CI of (−2.5 to +3.4), which is interpreted as no mean treatment differences existing in terms of FEV_1. The negative value does not cause a problem for the interpretation of the clinical effect (it just means that the treatment is worse).

However, cost-effectiveness ratios are not straightforward, and often point estimates (average) ratios are not close to the value 1 (equivalent to mean differences of zero). The ratios could be negative, as stated in the previous paragraph. The 95% CI for an ICER of £2000 of (−£6000 to £1500) for a new treatment can be interpreted as meaning that the treatment is either cheaper and more effective (−£6000) or more costly and more effective (£1500). The value of −£6000 does not tell us whether the denominator or the numerator is likely to be negative. The interpretation can be even more difficult if there are several possible treatment comparisons. If the ratio is in quadrant 2, one possible interpretation is that the new treatment could be cheaper and more effective, but it could also be more expensive and more effective by £1500 per unit of effect.

Moreover, it is also possible to have two cost-effectiveness ratios with the same values, both negative, but with completely opposite meanings. An ICER of −£200, for example, could lie in either Quadrant 2 or Quadrant 3. On the one hand, in Quadrant 2, the ICER tells us that the new treatment clearly dominates (cheaper and more effective); on the other hand, in Quadrant 3, it could be interpreted as being more expensive with worse effectiveness. One should therefore exercise caution when interpreting CIs for cost-effectiveness ratios.

For these reasons, among others, the INMB is a preferred way of presenting the results of economic evaluation and the uncertainty around the estimates.

The INMB (Stinnett & Mullahy, 1998) changes the scale of the cost-effectiveness ratio.

If

$$\frac{\Delta\mu_{A-S}}{\Delta\varepsilon_{A-S}} < 1$$

then $\Delta\mu_{A-S} < \Delta\varepsilon_{A-S} \times \lambda$ (simple algebraic manipulation)

$$\Delta\mu_{A-S} - \Delta\varepsilon_{A-S} \times \lambda < 0$$

Hence

$$\text{INMB} = \Delta\varepsilon_{A-S} \times \lambda - \Delta\mu_{A-S} > 0 \qquad (2.2)$$

Or, more simply, INMB = $\Delta_e \times \lambda - \Delta_c > 0$, using earlier notation.
As long as the INMB is >0, the new treatment is considered cost-effective.

Example 2.3: Computing the Incremental Net Monetary Benefit

If the difference in effects is 0.9 (Δ_e) and the difference in costs (Δ_c) is
£5,000, then for a given WTP of λ = £30,000, the INMB = 0.9 × £30,000 −
£5,000 = £22,000. Since the value of the INMB is >0, the new treatment is
considered to give a net benefit which is valued at £22,000. This is equivalent to the ICER being less than the value of λ. In this case, the value
of Y = $\Delta\mu_{A-S}/\Delta\varepsilon_{A-S}$ = Δ_c/Δ_e = £5000/0.9 = £5555. Clearly, this value is
$<\lambda$ = £30,000, the threshold, and hence both ICER and INMB approaches
agree that the new treatment is cost-effective at a value of λ = £30,000.

A ratio on its own is not interpreted as a net monetary benefit (NMB)
(the number is just a ratio of two quantities). The INMB approach avoids
the need for conducting statistical tests for a ratio. The INMB appears
to be the preferred choice for some countries to base a decision on
cost-effectiveness.

One useful way of considering the INMB is that the ratio is transformed
onto a linear scale (Figure 2.3). That is, Equation 2.2 can be presented as a
regression line:

$$\text{INMB} = \Delta_e \times \lambda - \Delta_c$$

$$Z = \Delta_e \times \lambda - \Delta_c$$

where Z is the INMB

$$= -\Delta_c + \Delta_e \times \lambda$$

This is known more familiarly as the equation of a straight line of the form
Y = mW + c, where Δ_c is the intercept (c), Δ_e is the slope and λ is the equivalent of W, which takes on different values. The INMB is on the y-axis and λ is
on the x-axis on the graph (see Figure 2.3).

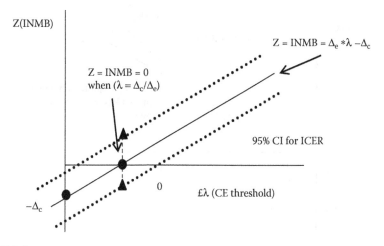

FIGURE 2.3
Cost-effectiveness using incremental net monetary benefit (INMB).

- Starting with $Z = -\Delta_c + \Delta_e^* \lambda$, when the value of $\lambda = 0$ (i.e. not prepared to pay anything for the new treatment), Z (the INMB) takes a value of $-\Delta_c$. This means that if one is not prepared to pay anything for a new treatment, the new treatment has no benefit (valued in pounds). In fact, this value is simply the mean difference in costs, $-\Delta\mu_{A-S}$, or a net monetary loss.

- When the payer is prepared to pay more for the new treatment, that is, as λ increases, the INMB also increases. For example, for a value of $\lambda = £20,000$ and $\Delta_e = 1$, the value of Z (INMB) is >0.

- The two solid triangles in Figure 2.3 represent the 95% CI for the ICER or cost-effectiveness ratio at the point where the INMB = 0. The dotted lines represent the 95% CIs for the INMB for different values of λ. When the 95% CIs (dotted lines) cross the y-axis, this gives us a 95% CI for the difference in mean costs.

- There is also a point where the INMB = 0. This is where the line $Z = -\Delta_c + \Delta_e^* \lambda$ crosses the x-axis, or when Z = 0. The value of λ at this point is where the new drug offers zero INMB, but also does not offer a monetary loss. It is also the point at which we get the value of the cost-effectiveness ratio $\lambda = \Delta_c/\Delta_e$.

Although Figure 2.3 presents the INMB versus λ, one could also present this as two lines, one for each treatment, as in Figure 2.4, where each line, for each treatment, is presented as a regression line, using slightly different notation, without loss of generality. In Equations 2.3 and 2.4, these are the NMBs (dropping the *incremental*):

$$NMB_A = E_A \times \lambda - C_A \qquad (2.3)$$

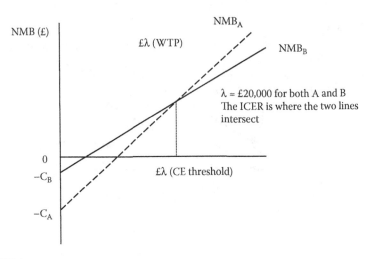

FIGURE 2.4
Net benefit chart for two treatments separately.

$$NMB_B = E_B \times \lambda - C_B \tag{2.4}$$

where E_A, E_B are the mean effects of A and B, respectively, and C_A, C_B are the mean costs for A and B, respectively.

In Figure 2.4, the NMBs and costs are plotted for each treatment separately. The intercepts on the y-axis of $-C_A$ and $-C_B$ represent the costs incurred for each treatment when the WTP (cost-effectiveness threshold) is zero. The difference between the intercepts is the incremental cost (Δ_c). The slope of each line (E_A, E_B) is a measure of effectiveness (e.g. QALYs or some other unit). Increasing costs are represented by the line shifting downwards and decreasing costs by the line shifting upwards.

As in most regression models, the steepness of the slope has an interpretation. In this example, the steeper slope (Treatment A) has greater effectiveness compared with Treatment B. At some point $\lambda = K$, the NMBs for Treatments A and B are equal (where the lines intersect). At values of $\lambda > K$, Treatment A gives higher values of NMB than Treatment B (Figure 2.4). In other words, the NMB at $\lambda = \lambda^*$ is equal to 0, and for $\lambda^* > 0$, the NMB is >0 (λ^* is any particular value of λ).

At the intersection of these two lines, the NMB for both treatments is equal:

$$NMB_A = NMB_B$$

so

$$E_A \times \lambda - C_A = E_B \times \lambda - C_B$$

Hence,

$$E_A \times \lambda - E_B \times \lambda = C_A - C_B$$

$$\lambda(E_A - E_B) = C_A - C_B$$

$$\lambda = \frac{(C_A - C_B)}{(E_A - E_B)}$$

which is the cost-effectiveness ratio.

Therefore, the intersection of the two lines (Figure 2.4) is also the cost-effectiveness ratio. Figure 2.4 makes the usefulness of the NMB approach clear. More information is displayed on a net benefit chart (as in Figures 2.3 and 2.4) compared with the cost-effectiveness plane. In the cost-effectiveness plane, the value of the ICER needs to be compared with various values of λ, and the number of points above or below the line is difficult to visualise for changing values of λ. In addition, for the graph of a single line (for the INMB), the cost-effectiveness ratio is simply where the line cuts the x-axis, and the 95% CI for the INMB at various values of λ can be presented. It is of interest to note that at the point where the INMB = 0, these limits also correspond to the so-called Fieller limits (Stinnett & Mullahy, 1998). If the 95% confidence bands in Figure 2.3 do not cut the x-axis, then, as noted earlier, problems with inference exist.

We now provide two examples using SAS for estimating the INMB. The SAS and STATA programs for estimating the Fieller 95% CIs and bootstrap estimate are shown in Appendix 2A.1 and 2A.2.

Example 2.4: Using SAS to Estimate the INMB

Using data from 20 patients from a Phase III patient-level clinical trial, we calculate the mean INMB.

The costs (Chapter 4) are taken from several sources in the study captured at the patient level. For example, Patient 1 had a side effect (a rash, which was treated with ointment at a cost of £20); there was also the cost of treatment at £500; and in addition, the other costs (e.g. treatment administration costs) totalled £200. The total cost for Patient 1 associated with Treatment B was therefore £20 + £500 + £200 = £720. The main point here is that the cost for each patient is obtained from different component costs added together.

For calculating the NMB, we use Equations 2.3 and 2.4 for each patient in each treatment group. We will assume that Treatment A is the new treatment and B is the standard. The NMB_A for Treatment A = $E_i \times \lambda - C_i$, where E_i is a measure of effectiveness ($i = 1$–12) for each patient (in this case a QALY) and C_i is the total cost of treating patients receiving Treatment A (or B). For example, for a value of $\lambda = £20,000$ (assumed in this example), the NMB_B for Patient 1 (Treatment B) is $2.1 \times £20,000 - 720 = £41,280$ (using data in Table 2.1).

TABLE 2.1

Data for Example 2.4

Patient	Treatment	Cost	QALY[a]
1	B	£720	2.1
2	B	£840	2.4
3	B	£620	3.1
4	A	£350	6.4
5	A	£280	7.2
6	B	£540	1.4
7	A	£590	7.3
8	A	£490	6.5
9	A	£280	5.2
10	A	£200	8.9
11	B	£1290	1.4
12	B	£820	0.9

[a] QALY calculations are explained in more detail in Chapter 5.

If the data are in an SAS data set called 'example 1', then the following SAS code could be used to model the observed NMB against treatment. We use a fixed effects model for now. A more complex model (such as including centre effects in the case of a multicentre trial) can be included, where centre effects are random (i.e. the centres from which patients were recruited in the trial are considered to be a random selection of all possible centres). The NMB calculation in the SAS data set 'example12' uses Equations 2.3 and 2.4 ; 'lambda' is λ, the effect E is termed 'qaly' and cost labelled 'cost' in the following SAS code:

```
Data example12;
Set example 1;
Lambda=20000
NMB=(lambda*qaly)- Cost;
Run;
**Fixed Effects model with treatment as the main effect and
response is
NMB***;
proc mixed data=example12;
class trt;
model nmb=trt / solution;
lsmeans trt / cl pdiff;
estimate 'A vs B' trt 1-1 /cl;
run;
```

This 'proc mixed' code generates:

1. The mean NMB for Treatments A and B
2. The INMB with the corresponding 95% CI

The model could also incorporate other factors, such as centres, to provide adjusted means (least squares means 'LSMeans') which control for

TABLE 2.2

Output from SAS Code to Estimate Mean INMB for Example 2.3

Estimate	Value	Lower 95% CI	Upper 95% CI
Intercept	36861.67		
A	101106.66 (INMB)	74333	127880
B	0.000		
Label	**LSMean**		
A	137968.3 (NMB$_A$)		
B	36861.7 (NMB$_B$)		

differences in centres (or other confounders). An example of a mixed effects model will be shown later using an example from Manca et al. (2005). The important part of the output for the analysis in Example 2.4 is shown in Table 2.2.

In Table 2.2, the mean (LSMean) NMB for each treatment group is £137,968.3 and £36,861.7 for A and B, respectively. The INMB is the difference, which is £101,106.67 in favour of Treatment A. The uncertainty around the INMB is given by the 95% CI (£74,333 to £127,880). Since the INMB is positive and the 95% CI does not contain zero, Treatment A is considered to be cost-effective compared with Treatment B (in terms of value for money).

The two NMB models can be written for the two treatments as

$$£137,968.3 = E_A \times £20,000 - C_A \tag{2.5}$$

$$£36,861.7 = E_B \times £20,000 - C_B \tag{2.6}$$

Although the inference for the INMB is determined from the SAS output, what is not clear is what the incremental cost (Δ_c) or effects (Δ_e) are. In Equations 2.5 and 2.6, there are four unknowns (E_A, E_B, C_A and C_B). The main objective in this example was to estimate the INMB. The approach to estimating the remaining quantities requires fitting a separate model for costs (regress total costs or against treatment to estimate mean costs for each treatment) and effects.

The INMB model is therefore

$$INMB = £101,106.63 \times Treatment$$

If treatment is coded as 1 for Treatment A, and 0 for B, then inserting these values in the model provides the NMB for Treatments A and B.

Example 2.5: Using SAS to Estimate the 95% CI for the ICER Using the Simple Bootstrap

Following Briggs (1997), Briggs et al. (1998) and Willan & Briggs (2006), we provide an estimate of the cost-effectiveness ratio using the bootstrap method. The SAS/STATA programs are found in Appendix 2A.1

and 2A.2. We will describe and interpret the output for the simple data set of Example 2.3 and briefly explain the SAS/STATA code.

In this example, we first estimate all mean costs and effects using least squares means (LSmeans) and not the raw means. The LSmeans may be different from the raw means, particularly in cases of missing data and imbalances in the number of patients per treatment group.

1. First generate 1000 bootstrap samples, where each bootstrap sample consists of 12 observations. A bootstrap sample randomly selects each patient's set of data (costs and effects) and replaces that patient before selecting another patient's data. It is possible that the same patient may be (randomly) selected again.
2. Once we have sampled 12 patients with their entire data, we then run a regression model to estimate the LSmean costs and effects for that bootstrap sample. Since we have 1000 bootstrap samples, we will have 1000 mean costs and effects for each of Treatments A and B.
3. For each bootstrap sample, we can compute the ICER using the formula introduced earlier: Δ_c/Δ_e. Therefore, we will have 1000 ICERs.
4. Now, from the 1000 ICERs we can generate the required bootstrap statistics. For our purposes, we present the bootstrap mean ICER (the mean of all 1000 bootstrap means) with the empirical quantiles. The quantiles are obtained by sorting the ICERs in order and reading off the values at the lower 2.5% and upper 97.5%.

Table 2.3 shows that the bootstrap mean ICER is £89.12/QALY (negative ratio, interpreted such that A is cheaper and more effective), which is comparable to the observed value of £87.4/QALY. The 2.5th percentile value is £138.7/QALY for the observed data. Table 2.3 shows the quantiles from the 1000 bootstrap ICERs. The data show considerable variability in the bootstrap estimates. However, we can be reasonably confident that the true ICER ranges from £138/QALY (cheaper on Treatment A) to only £45/QALY (cheaper) with about 95% confidence.

The SAS procedure 'SURVEYSELECT' is the key procedure which generates the 1000 bootstrap samples (Cassell, 2010).

Output from SAS showing Bootstrap Quantiles for the ICER for Example 2.4 is shown in Table 2.4.

TABLE 2.3

Output from SAS to Estimate Bootstrap Mean ICER

Bootstrap Statistics			
Mean	−89.12	Std Deviation	23.67
Median	−87.66	Variance	560.64
Mode	−119.67	Range	161.46
		Interquartile Range	31.19

TABLE 2.4

Output from SAS Showing Bootstrap Quantiles for ICER

Quantiles from Bootstrap									
0	1%	2.5%	5%	25%	50%	75%	95%	97.5%	99%
−183.7	−148.8	−138.7	−128.3	−104.1	−87.6	−72.9	−52.4	−45.01	−38.9

As indicated earlier, Fieller's theorem is used to estimate the uncertainty of a ratio quantity, in particular the ICER. CIs using Fieller's theorem are still reported for the assessment of cost-effectiveness (Gregorio et al., 2007; Carayanni et al., 2011, among others). Fieller's theorem has also been used in other applications, such as bioassay and bioequivalence trials (Chow & Liu, 2000). Use of Fieller's theorem allows the calculation of an exact CI for the ratio of two means without doing any kind of transformations to the data. We assume that the ratio of the means (mean difference in costs divided by mean difference of effects) follows a bivariate normal distribution (e.g. Briggs 1997; Willan & Briggs 2006; Chow & Liu 2000) in order to use this method, which may be a tenuous assumption, because first, the denominator has a good chance of being zero, as in the case of equivalence and non-inferiority-type trials, and second, the costs may not always satisfy the conditions for normality (Mihaylova, 2011). Further details of issues surrounding the CIs for the ICER using Fieller's theorem can be found in Willan & Briggs (2006). Siani et al. (2003) suggest that Fieller's theorem is robust under many conditions, should be used all the time, whatever the form of the confidence region, and may not be the same as the bootstrap estimate. Indeed, criticisms of the Fieller CI for the ICER are considered invalid (Siani et al., 2003).

Example 2.6: Estimating the 95% CI for the ICER Using Fieller's Theorem

The technical aspects of Fieller's theorem, which can be complicated, are provided in Appendix 2B.1. No SAS or STATA code is really needed, because the CIs can be computed by inserting relevant values of mean costs and effects with their variabilities in the equations in Appendix 2B.1 (which can also be computed using an Excel spreadsheet). It is more important for the user to know how to interpret the CIs.

Table 2.5 shows the results of the inputs into the equation in Appendix 2B.1. The procedure IML in SAS can also be used to compute the confidence intervals using Fieller's theorem (see Stokes & Koch, 2009).

Table 2.5 shows how the 95% CIs for the ICER are close to the bootstrap estimates, confirming the earlier conclusion that the new treatment (Treatment A) is cheaper and more effective (the negative values should not be taken to mean that the effect is worse on Treatment A).

TABLE 2.5

Output from SAS Showing Mean ICER and Confidence
Intervals Using Fieller's Theorem

Mean ICER	Lower 95% CI	Upper 95% CI
−89.10	−137.4	−43.2

2.3 The Concept of Dominance

The concept of dominance is important in economic evaluation, especially
when there are more than two treatments being compared. In health economic evaluation, a treatment is 'dominated' if there is an alternative treatment that is less costly and more effective. Dominance can be demonstrated
using the ICER or the INMB, but as was noted in Section 2.2, negative ICERs
make it difficult to work out which treatments are dominant. The INMB,
on the other hand, provides consistent ordering, and a decision as to which
treatment dominates (or is dominated) is based on the largest INMB.

Example 2.7: Dominance

In this example (Table 2.6), Treatment S is dominated by Treatment A
(A vs. S) because the incremental cost is −£10,000 (£15,000 − £5,000) and
the incremental effect is 4 (8 − 4). Treatment A has, therefore, strictly
dominated S because it is less costly and more effective (Gray et al., 2011).
It is worth pointing out again that reporting a negative ICER as merely
−£2,500 for A versus S does not tell us which part (numerator or denominator) is negative and which is positive. It is difficult to know where in
the cost-effectiveness plane the ICER lies. Therefore, reporting incremental costs and effects clarifies this.

The term *extended dominance* refers to a new treatment being more costly,
more effective and providing greater value for money in the form of a lower
ICER. When extended dominance occurs, a Treatment X (Table 2.7) may

TABLE 2.6

New Treatments (A, B and C) Compared with Standard Treatment (S)

	A	B	C	S (Standard)
Cost (£)	5,000	50,000	25,000	15,000
Effect (QALY)	8	7	9	4
Incremental Cost* (£)	−10,000	+35,000	+10,000	
Incremental Effect*	+4	+3	+5	
ICER	−2,500	11,666	2,000	

*A−S, B−S, C−S (treatment minus standard).

TABLE 2.7

Extended Dominance for Three Treatments X, Y and Z

	Total Cost	Effects	Comparison	Incremental Cost	Incremental Effects	ICER
X	£27,000	4.5	Y vs. X	£11,000	1.5	£7333
Y	£38,000	6	Z vs. Y	£14,000	3	£4666
Z	£52,000	9	Z vs. X	£25,000	3.5	£7143

appear to be more effective and costly than another treatment (as with X and Y in Table 2.7). However, it is possible that by using a combination of Treatments X and Z together, less cost and more effectiveness are obtained, so that Y is dominated. This is called extended dominance.

Example 2.8: Extended Dominance

Sort treatments in order according to cost, from low to high. In this example, the order is £27,000, £38,000 and £52,000 for X, Y and Z, respectively.

1. Calculate the incremental cost and effects for X, Y and Z by comparing each treatment with the treatment above it. For example, Y is compared with X, and Z is compared with Y. We therefore have three incremental costs and effects: Y versus X, Z versus Y and Z versus X. These are: £11,000 and 1.5 QALY for Y versus X; £14,000 and 3 QALY for Z versus Y; and £25,000 and 3.5 QALY for Z versus X.
2. Extended dominance may not always be possible. In practice, it is not possible or practicable to simply combine treatment options for cost-effectiveness. Hence, some authors (Postma et al., 2008) consider the idea of extended dominance as 'theoretical' and difficult in practice, citing the case of a strategy of screening patients for sexually transmitted diseases as an example. Screening all women and just 40% of men, for example (a combination strategy), might be difficult to implement in practice for various social and political reasons such as inequality issues, despite the combination being the most cost-effective.

2.4 Types of Economic Evaluation

In this section, the key methods for performing an economic evaluation are discussed. Some of these have already been introduced briefly, and we now define them formally and summarise their use in Table 2.8. Cost–utility analysis (CUA) and cost-effectiveness analysis (CEA) are the more common methods found in the literature.

TABLE 2.8

Cost–Utility Analysis

Treatment	Erlotinib	Docetaxel	Difference	ICER[a]
Total costs (£)	13,730	13,956	−226	
Effects				
PFS (months)	4.11	3.33		
Post-progression survival (months)	4.92	5.56		
OS (months)	9.03	9.03		
Utility (before progression)	0.426	0.451		
Utility (after progression)	0.217	0.217		
Total QALY	0.238	0.206	0.032	−£7,106

Source: Lewis et al., *J. Int. Med. Res.*, 38, 9–21, 2010. With permission.
Note: OS: overall survival; PFS: progression-free survival.
[a] ICER calculated as −£226/0.032.

2.4.1 CUA

CUA is a method of economic evaluation which compares differences in mean costs with differences in mean utilities. Utility in this context means patient preference. The most common measure of utility in economic evaluation is the QALY. Others include health years gained (HYG) and disability-adjusted life years (DALYs). For now, it is sufficient to say that in practice, eliciting patient preference in a clinical trial directly (for a particular treatment or health condition) is often difficult. Research ethics committees (RECs) might object to complex questioning of patients to select preferences for particular health conditions (health states).

Utilities quantify preference for treatments expressed through generic QoL assessments (such as the EuroQol EQ-5D). CUA captures duration of disease (e.g. time to disease progression), toxicity, QoL and other measures into a single composite measure such as QALY or DALY. The advantage of the CUA is that one can compare the cost per QALY between several different disease areas. For example, the cost per QALY in cancer patients can be compared with the cost per QALY in diabetes patients. In general, the lower the cost/QALY, the better (i.e. the more cost-effective the treatment is deemed to be).

CUA is often found in clinical trials for diseases such as cancer, in which quantity of life and QoL are important. If the relevant QoL measure cannot be captured in the clinical trial, external data can be sought to determine QALYs. Outcomes in a CUA are often reported in terms of cost per QALY.

Example 2.9: Cost-Effectiveness of Erlotinib versus Docetaxel

The cost-effectiveness of Erlotinib versus Docetaxel (Lewis et al., 2010) compares the value of these treatments for non-small cell lung cancer patients. Table 2.8 summarises the costs, effects and estimates of QALYs. In this trial, the utilities were not collected prospectively, but a separate *utility study* was undertaken in a healthy population (a utility study is a separate observational study or survey which collects data on preferences for particular health states). Note also how utility is reported by pre- and post-disease progression. In this example, the efficacy results were taken from published clinical trial data (which made several assumptions), but utilities were taken from elsewhere.

The conclusion from Table 2.8 is that the treatment Erlotinib was cheaper and more effective than Docetaxel (hence the negative ICER).

2.4.2 CEA

CEA is used to compare the costs and effects of two or more treatments in terms of a natural or common effect (e.g. cost per subject who has quit smoking using a new smoking cessation device, cost per case of measles). The costs can also be expressed in terms of the number of life years gained (a final outcome) as well as the intermediary outcomes, such as cost per quitter or per case of measles. The life years gained in this example can be estimated from life tables. For example, a 55-year-old with type 2 diabetes who smokes and has a systolic blood pressure (SBP) of 180 mmHg, a total HDL cholesterol ratio of 8 and HbA1c of 10% might have a life expectancy of about 13.2 years, 5 years after diagnosis; similarly, for a patient who is a non-smoker with SBP of 120 mmHg, total/HDL ratio of 4 and HbA1c of 6%, a life expectancy of 21 years is expected 5 years after diagnosis (Leal et al., 2009). Based on each patient's age and diagnosis, the estimated life expectancy, and hence life years gained, is determined and then adjusted with the QoL (utility) data generated from the trial (or elsewhere).

Example 2.10 : Cost-Effectiveness of Larval versus Gel Ointment for Leg Ulcers

An example of CEA is adapted from Soares et al. (2009), in which leg ulcers are treated by larval therapy or a gel ointment (standard of care). The primary end point is time to healing (TTH). The larval therapy is on average £1.11/day more expensive than the ointment (see Table 2.9).

TABLE 2.9

Cost-Effectiveness Analysis

Treatment	Larval Therapy	Gel (Standard)	Difference	ICER[a]
Total costs £	2073	1976	97	
Efficacy				
TTH (days)	120	207	87	
Cost/day £	17	9.50	7.50	1.11/day

Source: Soares et al., *BMJ*, 338, b825, 2009. With permission.

[a] Calculated as £97/87 days

2.4.3 Cost–Benefit Analysis (CBA)

CBA is a method of economic evaluation in which monetary values are placed on both the inputs (costs) and outcomes (benefits) of healthcare. Final results are reported in monetary values, and therefore different treatment indications are compared using a common unit of measurement (money). Placing a monetary value on all effects can be controversial. For example, valuing 6 months of survival time in terms of money may be considered unethical, even though many people take out life insurance (which also places an upper limit of compensation on early death). There are various methods which discuss how people (patients) may place value on health benefits in monetary terms, through the idea of asking how much they are 'willing to pay' (Gray et al., 2011; Robinson, 1993), but these are unlikely to be encountered in clinical trials in practice because of the difficulty in translating clinical effects to monetary value, although in some trials of prevention, a CBA may be more appropriate (Hornberger, 1998).

2.4.4 Cost-Minimisation Analysis

Cost-minimisation analysis (CMA) is used when the case for a new intervention has been established and the treatments under consideration are expected to have the same or similar outcomes. Similar outcomes are often important for equivalence or non-inferiority-type trials. In these circumstances, attention may focus on the numerator of the ICER to identify the least costly option (Rascati, 2009; Robinson, 1993). The results of a CMA are reported with greater emphasis on differences in mean costs. However, as discussed earlier, it is more appropriate to consider the costs and effects together (the joint distribution), regardless of equivalence in effects. This would be best seen by reporting the INMB and not the ICER, which is likely to have a large value (because the denominator is so small).

> **Example 2.11: Cost-Minimization Analysis of Laparoscopic versus Small-Incision Cholecystectomy**
>
> Table 2.10 summarises a CMA from a randomised controlled trial in patients diagnosed with symptomatic cholecystolithiasis who were randomised to either laparoscopic or small-incision cholecystectomy (Keus, Trials 2009). The outcomes are not statistically different, with all p-values >.05. There are, however, statistical differences in mean costs. The INMB for the complications would be 5% × £20,000 − £545 = £455 in favour of laparoscopic surgery at a cost-effectiveness threshold of £20,000. At a threshold of £30,000, the INMB is £955. The INMB was not reported in the original publication.

Table 2.11 summarises the different types of economic evaluation.

TABLE 2.10

Cost-Minimisation Analysis

Treatment	Laparoscopic (N = 120)	Small Incision (N = 137)	Difference	Conclusion[a]
Direct costs £	305,760	452,654		
Direct costs/patient £	2,548	3,304	−756	p = .006
Indirect costs £	133,406	123,404		
Indirect costs/patient £	1,112	901	211	p <.001
Total costs/patient	3,660	4,205	545	
Effects				
Mortality rate (%)	0	0	0	
Complications (%)	21	16	5	p = .11
Intra-operative (%)	5	3	2	p = .36
Post-operative (%)	16	13	3	p = .33

Source: Keus et al., *Trials*, 10, 80, 2009.

[a] Conclusion based on p-values for differences in mean costs.

2.5 Statistical versus Health Economic Models

The word 'model' can have a variety of different interpretations. A model can be as simple as a map of England. It provides some abstraction of reality. In statistics and mathematics, a model is typically an equation which attempts to explain the relationship between observed outcomes (response variables) and some underlying 'causal' factors. For example, a response outcome labelled Y could be collected over time (e.g. 24 weeks), as in a clinical trial; factors (labelled X) such as age, weight and compliance rates might be assumed to be predictive of Y. The relationship might be expressed as a mathematical equation similar to the relationship between INMB and λ (the cost-effectiveness threshold) discussed in Section 2.1. The mathematical relationship might be linear or non-linear. One purpose of such a model might be to simply determine whether X 'predicts' or in some way influences the response Y with some degree of certainty (such as the equation of a straight line: $Y = mX + c$).

A model may also be used to predict values of Y measured beyond 24 weeks, known as extrapolation, because the model is used to predict values outside the observed data range. In cancer trials, survival models are used to predict probabilities of survival outside the clinical trial observation window so that a lifetime QALY can be computed. For example, in certain types of cancer (e.g. lymphoma), the overall survival of patients could be >10 years. It may be prohibitively expensive to follow up patients for 10 years in a clinical trial. The trial may have only a 4 year observation window, enough to observe most patients up to disease progression. A model might be used to

TABLE 2.11

Summary of Types of Economic Evaluation (Comparing New versus Standard Treatments)

Method of Economic Evaluation	Costs Measurement	Example Clinical End Points	Example Effectiveness Measures	Conclusions Best Reported in Terms of	Clinical Objective
CUA	Currency (£, $, €)	Overall survival, PFS, rates of infection	Utility (QALYs, DALYs), QoL	ICER, INMB	Superiority, non-inferiority
CEA	Currency (£, $, €)	Blood pressure, quit rates (e.g. smoking), cure rates	Natural units, LYG, life years gained; LYS, life years saved	Cost per unit, INMB	Superiority, non-inferiority
CMA	Currency (£, $, €)	Any (clinical differences not expected)	Any	(£, $, €>), INMB	Non-inferiority
CBA	Currency (£, $, €)	Any	Any	(£, $, €)	Resource allocation

predict the proportion of patients alive after 4 years or even to estimate the survival probabilities from elsewhere (e.g. from published life tables).

The type of statistical model which involves the use of patient-level data may not be possible for some types of health economic evaluation models, often because patient-level data is not always available. Aggregate data (e.g. summarised/synthesised from clinical trial reports or publications) form a basis for populating some health economic evaluation models. For example, the proportion alive 5 years after taking a new treatment might be calculated in several ways. One way is by taking the average (mean) of 5 year survival rates reported across several published trials. If most of the published data report 1 year survival data, then this estimate would be unreliable, and therefore an alternative approach to estimating 5 year survival data is needed for a health economic evaluation. Another way might be to model the patient-level survival times and extrapolate or predict 5 years survival using a statistical (parametric) survival model. When patient-level survival data are not available, the alternative approach is to use a Markov-type model which just depends on knowing the initial survival rate at 1 year and then extrapolating (with sometimes unjustifiable assumptions) future survival rates. Barton et al. (2004) discuss the features of statistical and other models used in economic evaluation.

It is often useful to ask what we need from the chosen model. In a statistical model, we might need the parameter estimates (hazard rate, odds ratios), CIs, p-values and so on. With such models, final conclusions are based on some type of average measure. In economic evaluation, final reporting is almost always made in terms of the expected value (i.e. the mean) with the uncertainty around it expressed. There is some overlap between a purely statistical model and a health economic model.

Rittenhouse (1996) describes two particular types of model which concern us in the context of patient-level data:

a. Deterministic models
b. Stochastic models

2.5.1 Deterministic Model

A deterministic model makes statements with a degree of certainty (Rittenhouse, 1996). For example, a patient who is feeling unwell might:

(i) Go to the doctor (decision)
(ii) Get prescribed medicine (treatment allocation)
(iii) Take the treatment (compliance)
(iv) Become better (outcome)

In steps (i)–(iv), there is no probability element associated with each step/path: it is assumed that patients will get better if they take a medicine.

2.5.2 Statistical Model

A stochastic model, on the other hand, might consider the *likelihood* of getting better. Each of the steps (i) to (iv) could have been made stochastic by attaching probabilities (chance of going to the doctor, chance of taking the treatment, chance of getting better). The statistical models (e.g. regression type models) are also examples of stochastic models, because they assume that the observed outcomes and their expected values (i.e. mean values) follow a probability distribution.

In an economic evaluation alongside a clinical trial, in which patient-level data are collected, a stochastic model is considered appropriate. The main types of models discussed in this book are decision tree, Markov, cohort and simulation models (Chapter 6).

Statistical models are commonly based on patient-level data, while some health economic models use summarised data from clinical trial reports as inputs (inputs are often mean values of costs and effects) for Markov or decision tree models. Willan & Briggs (2006) discuss statistical models for patient-level costs to derive mean estimates of costs. There is sometimes confusion over which model to use in the presence of individual-level data. For example, if patient-level data are available from a clinical trial, what need is there to use summaries (e.g. of response rates or means) as inputs in a Markov model? The answer to this question lies in answering a separate question: which model is best for extrapolation? We recall that the clinical trial has a limited observation window. For example, if 25% of patients were infection free at 24 weeks, a Markov model could be used to 'extrapolate' not only the proportion with infection at 52 weeks, but also how long it would take for zero patients to remain alive or without an infection (or other health state), hence, the Markov model can provide estimates over a lifetime horizon. A statistical (stochastic) model for this approach might be complicated, as it involves explicitly modelling numbers of patients who move between different health states (e.g. infection, non-infection, death).

Brenan et al. (2006) provide a detailed summary of the models used in economic evaluation. The German Institute for Quality and Efficiency in Healthcare (IQWiG) guidelines (2008) also provide guidance on approaches to modelling. Often, little distinction is made between the statistical models and the health economic approaches to modelling. Both types of model assume that responses satisfy the independence assumptions, unlike epidemiological models, in which an interaction can occur due to one individual infecting another (Barton et al., 2004). We simplify the presentation of the models in Brenan et al. (2006) in Figure 2.5.

In summary, the statistical models that are used in clinical trials for demonstrating efficacy with patient-level data are not used for extrapolating data beyond the observation/follow-up period of the trial; statistical models need not provide estimates of treatment effects based only on the mean values

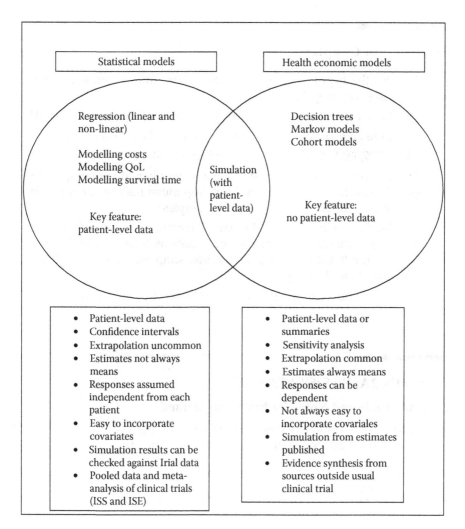

FIGURE 2.5
Statistical models and models used in economic evaluation.

(e.g. in some cases medians or geometric means can be used as a measure of treatment effect) and involve reporting uncertainty in terms of confidence intervals. Moreover, statistical models used in clinical trials do not necessarily depend on data synthesised across trials or other sources. Health economic models, on the other hand, may not involve using patient-level data; summary statistics are often used as inputs, and estimates of treatment effects are almost always reported in terms of the mean value; uncertainty is expressed through sensitivity analyses and simulation (Chapter 7), not CIs. Data analysis can involve using published data, and the extrapolation of data beyond the observation window is not uncommon.

Exercises for Chapter 2

1. Discuss the types of economic evaluations undertaken and when each one is appropriate.
2. What is the difference between an economic and a statistical model?
3. Compute an estimate of a bootstrap value for the following values:

   £234, £127, £698, £135, £875, £111, £985, £667, £346, £999, £126

   Discuss your choice of sample size and number of bootstrap samples. Would the estimate of the bootstrap mean make a difference if you double the number of bootstrap samples?
4. For the data in Table 2.1, if the cost of Treatment B was 30% higher with a 10% increase in QALY for each patient, what would the estimate of the ICER be? Compute the bootstrap estimate of the ICER with the 5% and 95% quantile values.
5. Why is extended dominance important from a payer perspective?

Appendix 2A: SAS/STATA Code

2A.1 SAS Code and Notes for Bootstrap Method

```
**procedure that carries out simple bootstrap**;
**If the data set contains for each subject the following
variables:
*Total costs
*Total effects
*Treatment Group
*Patient ID

*The procedure 'survey select' shown below can be used to;

*generate 1000 bootstrap samples and each bootstrap sample is;

*the same size as the data set. For example if the data set
was a sample size of 80, 1000 data sets of a sample size of
80;

*would be generated with the above variables;

proc surveyselect data=ds1 out=outboot
seed=30459584
method=urs
samprate=1
```

```
outhits
rep=1000;
run;

data boot1;
set outboot;
rename replicate=bootno;
run;
```

replicate is renamed as the bootstrap sample id

Now we need to get estimates of the model-based means for costs and effect. So we can use a mixed model for this.

```
ods output lsmeans=lsmcosts ;

proc mixed data=boot1 ;
by bootno;
class trt;
model cost=trt;
lsmeans trt;
run;

data lsmcosts2 (keep=trt estimate bootno);
set lsmcosts; **get least squares mean costs**;
run;

proc transpose data=lsmcosts2 out=lsmcosts3;
by bootno;
id trt;
var estimate;
run;

data bootdiffcosts;
set lsmcosts3;
incr_costs_boot=a-b;**get mean incremental costs**
run;
```

Get the bootstrap mean effects (QALY's);

```
ods output lsmeans=lsmqaly ;

proc mixed data=boot1;
by bootno;
class trt;
model qaly=trt;
lsmeans trt;
run;

data lsmqaly2 (keep=trt estimate bootno);
```

```
set lsmqaly;
run;

proc transpose data=lsmqaly2 out=lsmqaly3;
by bootno;
id trt;
var estimate;
run;

data bootdiffqaly;
set lsmqaly3;
incr_qaly_boot=a-b;**get mean incremental effects**;
run;

**merge bootstrap costs and qalys**;

data boot_all1;
merge bootdiffqaly bootdiffcosts;
by bootno;
icer=incr_costs_boot/incr_qaly_boot;**compute ICER**;
run;

**Sort ICER in order**;

proc sort data=boot_all1;
by icer;
run;

**statistics**;

proc univariate data=boot_all1;
var icer;
output out=final pctlpts=2.5, 97.5 pctlpre=ci;
run;
```

2A.2 STATA Command for Bootstrap

The STATA command for Bootstrap can also be accessed via http://www.uphs.upenn.edu/dgimhsr/stat-cicer.htm.

First compute the NMB for each treatment. Assign a variable 'Treat' for Treatment A or B (assume two treatments).

Then regress the NMB as a function of treatment: that is,

```
NMB = intercept +b*Treatment
```

In STATA this is

```
regress NMB treatment
```

To bootstrap the incremental INMB (the coefficient b), the STATA code is

```
regress NMB treatment
        bootstrap, reps(1000) seed(576): regress NMB
```

This will generate 1000 bootstrap samples for the INMB (the difference between two treatments).

Appendix 2B.1 Technical Details for Fieller's Method

If costs and effects (mean costs and effects) have a bivariate normal distribution, with true mean incremental costs $\Delta\mu_c$ and effects $\Delta\mu_e$, then Fieller's method can be used to compute the 95% (or other $Z\alpha/2$ confidence interval for the ICER = $\Delta\mu_c/\Delta\mu_e$). No SAS or STATA code is provided because there is an equation (a bit tedious), which simply requires entering the inputs. STATA code can be found at http://www.uphs.upenn.edu/dgimhsr/stat-cicer.htm, and an SAS code can be found in Faries et al (2010) and Stokes et al. (2010).

The Fieller equation for computing the CIs is

$$\frac{X_1 X_2 - k_{1-\alpha}\gamma_{12} \pm \sqrt{\left(k_{1-\alpha}\gamma_{12} - X_1 X_2\right)^2 - \left(X_2^2 - k_{1-\alpha 1}\gamma_2^2\right)\left(X_1^2 - k_{1-\alpha}\gamma_1^2\right)}}{X_2^2 - k_{1-\alpha 1}\gamma_2^2}$$

where:
   X_1 is the mean *difference* in cost, Δ_c (mean incremental cost)
   X_2 is the mean *difference* in effects, Δ_e (mean incremental effect)

γ_1^2 is the variance of the costs for Treatment A plus the variance of the costs for Treatment B (n_1 and n_2 are the sample sizes in each group):

$$\frac{\sigma_{CA}^2}{n_1} + \frac{\sigma_{CB}^2}{n_2}$$

γ_1^2 the variance of the effects for Treatment A plus the variance of effects for Treatment B (n_1 and n_2 are the sample sizes in each group):

$$\frac{\sigma_{EA}^2}{n_1} + \frac{\sigma_{EB}^2}{n_2}$$

and

$$\gamma_{12} = \frac{\rho_1 \sigma_{CA}\sigma_{EA}}{n_1} + \frac{\rho_2 \sigma_{CB}\sigma_{EB}}{n_1}$$

where:

ρ_1 is the correlation between costs and effects in Treatment Group A
ρ_2 is the correlation between costs and effects in Treatment Group B
σ_{CA} is the standard deviation of costs in Group A
σ_{CB} is the standard deviation of costs in Group B
σ_{EA} is the standard deviation of effects in Group A
σ_{EB} is the standard deviation of effects in Group B
$k_{1-\alpha}$ is the quantile of the normal distribution (when n_1 and n_2 are large)

The observed correlation can be easily computed in SAS using the PROC CORR option, the remaining parameters using the PROC Means, and the estimates inserted into the "Fieller Equation".

3

Designing Cost-Effectiveness into a Clinical Trial

3.1 Reasons for Collecting Economic Data in a Clinical Trial

There are several reasons for designing the cost-effectiveness component and collecting health economic resource data in a clinical trial. First, it is often cheaper to collect such data alongside a clinical trial than to carry out retrospective studies of costs and effects (Drummond, 1990). However, in some cancer trials, in which the follow-up period can be many years, it may be expensive to follow patients for extensive periods of time to collect relevant data. Delaying publication of important clinical findings on the efficacy of a new treatment in order to collect 'real-world evidence' on health resource use may raise some eyebrows. The economic value of a new treatment, however important, should not compromise medical knowledge and treatment benefits to waiting patients.

Second, several reimbursement agencies require economic data to be collected alongside clinical trials to demonstrate the value of new treatments. With many new treatments coming on to the market showing 'small' or 'modest' treatment effects, the burden of proof is on the claimant to demonstrate why any healthcare system should finance one treatment over the other. In the United Kingdom, the National Institute for Health and Care Excellence (NICE) makes specific reference to clinical trial data being used in assessing cost-effectiveness (NICE Guidelines, NICE DSU document 14 2011 and NICE 2013). When government agencies make specific requests, especially concerning payments for the provision of new drugs, these are not to be taken lightly which alone might justify why designing the economic component carefully in a clinical trial is important.

In Chapter 1, it was explained how an early decision to reimburse a new treatment can influence future revenue. Reimbursement authorities may request economic evaluation of new treatments closer to (and sometimes well before) a decision being made for licensing. Some reimbursement agencies may prefer to have an early review of the value argument (submission dossier) submitted. This may be in the interest of both the sponsor and the patients. Any delay in a recommendation from reimbursement authorities can unnecessarily restrict access to treatments by patients and healthcare professionals despite receiving market authorisation. The situation is not entirely the same in Germany, where a period of free pricing is offered (i.e. the pharmaceutical company gets an agreed price for the new treatment, while value is assessed).

In academic clinical trials, for which licensing is not relevant, reasons for collecting economic data are associated with evaluating the impact of the proposed health technology from a national health resource and government policy perspective. In some grant application forms, there is a desire to know from the perspective of the funder (e.g. National Institute for Health Research [NIHR], Clinical Trials Awards Advisory Committee [CTAAC], Health Technology Assessment [HTA]) whether the planned trial will impact future national healthcare resource use. The new health technology proposed need not be a new drug. For example, in a grant application, a new treatment might propose to compare following up patients intensively for tumour progression (e.g. $\geq$4 times a month) with a less intensive follow-up schedule (<4 times a month). The main objective might be to determine whether intensive follow-up results in earlier detection of disease progression. More intensive follow-up may lead to more efficient resource use in the long term compared with less intensive follow-up. Only follow-up schedules, not treatments, are compared in this type of trial.

A further reason for collecting economic (health resource) data in a clinical trial is because clinical trials provide strong internal validity of estimates of average costs and effects, but lack external validity, and therefore, additional evidence is sometimes needed to make informed decisions about the cost-effectiveness of new treatments (Sculpher et al., 2006) particularly in a real-world setting. If a decision cannot be reached regarding the cost-effectiveness of a new treatment based on data from one or more clinical trials, some trial-lists may question the reason for collecting economic data in the clinical trial in the first place. If the 'weight of evidence' from clinical trial data is considered too 'weak' to offer a robust conclusion for cost-effectiveness, then what economic data collected alongside a clinical trial adds to demonstrating the value of a new treatment becomes debatable. Other methods, such as value of information (VOI) and collecting additional data, might be useful when the total evidence available is not considered sufficient for a decision to be made. At best, one should consider that the economic data collected alongside a clinical trial may be answering a different question when it is insufficient to meet payer concerns. Gherghe et al. (2013) suggest operationalising the idea of generalisability and incorporating it into the design of trials, such as trials designed for the real-world setting with a view to getting market authorisation.

On the other hand, if external data (i.e. data outside the clinical trial) can be used for an economic evaluation to determine cost-effectiveness, then is there really any need to collect cost and outcome data prospectively in a clinical trial, if external validity is so important? Does mixing data from controlled and non-controlled sources provide unbiased estimates of treatment effects? For example, if we wish to carry out an economic evaluation comparing a new treatment with a standard, is it not possible to simply estimate the costs and quality of life (QoL) from published sources and use this information in combination with clinical trial data (efficacy data) as inputs into an economic model? Costs and effects from randomised and non-randomised trials may be handled separately (but the non-randomised evidence may be pooled with the randomised evidence for sensitivity analyses in some cases).

The answer to some of these questions depends on the magnitude of the differences between clinical trial data and non-clinical trial data. Kunz et al. (2008) conclude

> On average, non-randomised trials and randomised trials with inade-
> quate concealment of allocation tend to result in larger estimates of effect
> than randomised trials with adequately concealed allocation. However,
> it is not generally possible to predict the magnitude, or even the direc-
> tion, of possible selection biases and consequent distortions of treatment
> effects.

There are also differences in the way that costs (resource use) are collected in a non-randomised study compared with a clinical trial (Hltaky, 2002). For example, side effects on a placebo arm in a randomised trial may not be real costs in practice. It is not unusual to collect data on resource use that are of highest monetary value (bias) or that are likely to show differences between treatment groups. It is unlikely that every item of health resource use in a trial will be captured, and hence data collection forms (DCFs) or case report forms (CRFs) are often designed to collect limited health resource use data.

One important reason for collecting economic data in a clinical trial is simply because it is considered unethical to conduct a trial purely for the purposes of demonstrating value for money (although arguably, it is also unethical to waste resource when it could be put to better use in a more needy patient population). If bias is to be minimised, then the randomised controlled trial (RCT) framework may still be the only framework that accommodates *both* efficacy and effectiveness. There is some debate as to whether one should model (predict) the unknown or whether one should measure it in a clinical trial (NICE Guideline, 2013). The idea of a balance between internal and external validity was proposed earlier (Drummond, 1990). More than 20 years later, the argument has somewhat shifted from the idea that a pragmatic clinical trial with minimal inclusion/exclusion might be acceptable for showing cost-effectiveness to a situation in which a single clinical trial may no longer be admissible for demonstrating the value argu-ment (e.g. Sculpher, 2006). Combining clinical trial data with 'other' evidence to demonstrate value using complex methods (with sometimes unrealis-tic assumptions) appears to be the direction for showing the value of new treatments. If researchers are led to believe that the RCT data do not have 'enough' external validity for a conclusive reimbursement decision, then the impetus for collecting economic data may be lost in the trial. It would remain important to design the trial for cost-effectiveness in addition to efficacy, regardless of the potential payer decision, to postpone judgement until sufficient data are available. Ramsay et al. (2015) offer useful guidelines for trial design for cost-effectiveness, some of which are now discussed.

3.2 Planning a Health Economic Evaluation in a Clinical Trial

One key aspect of planning a health economic evaluation is to identify:

1. What data will be collected for economic evaluation?
2. How will the data be collected?
3. When will the data be collected?

The answers to these three questions are closely related to the overall objectives and the design of the trial, in particular whether the economic evaluation is prospective or retrospective. Whereas in almost all cases clinical trial data are collected prospectively, this is not always the case for cost and QoL data. Some possibilities are summarised in Table 3.1, which outlines the type of economic evaluation that could be undertaken (last column).

3.2.1 Costs/Resource Use and Effects Collected Prospectively

In the context of a clinical trial, costs and resource data are collected at the patient level prospectively. For each patient, there are several measures of effects and several measures of resource use (such as hospital visits, consultations, scans, number of doses, etc.). The trial is typically designed for a clinical end point, but the economic component is accommodated. In some cases, both clinical and economic end points are formally designed (including sample size calculations) into the trial. For example, sample size is likely

TABLE 3.1

Collecting Data in a Trial for an Economic Evaluation

Costs Collected	Clinical End Point (Primary)	Other Outcomes (e.g. Utility)	Possible Economic Evaluation Method
Prospectively	Prospectively	Prospectively	Stochastic (patient level)
Retrospectively	Prospectively	Prospectively	Stochastic/decision tree/Markov
Retrospectively	Prospectively	Retrospectively	Decision tree/Markov
Retrospectively	Retrospectively	Retrospectively	Decision tree/Markov

to be based on the clinical end point, as will timing of the main primary and secondary end points. However, the CRFs might be specifically designed to capture resource use and specific adverse events in detail, especially those associated with important costs. Extended patient follow-up time may allow the collection of data in the form of an open label extension with relaxed inclusion criteria (in a trial without an economic component, the follow-up period might be restricted).

The economic evaluation is usually performed after the results of the clinical effects are made available (although the planning of economic models will be much earlier). There is unlikely to be any benefit in carrying out an economic evaluation if there is no clinical benefit. The economic evaluation is usually carried out using stochastic methods; that is, using patient-level observations for the statistical analysis of costs and effects. Another approach to analysis is to summarise the individual cost and efficacy data (e.g. means and frequencies) and use these as 'inputs' into a health economic model (e.g. Markov or decision tree model) to estimate the incremental cost-effectiveness ratio (ICER).

3.2.2 Clinical Effects Are Collected Prospectively but Some Health Resource and Quality-of-Life Data Are Collected Retrospectively

The clinical trial in this scenario is designed only for the clinical component, and costs and utility (QoL) are extracted from external sources or published literature. For example, in an oncology trial, the clinical effects might be a measure such as progression-free survival (PFS) time – calculated from the time from randomisation until disease progression. The health resources involved in treating with the new drug, along with QoL data, are extracted from the clinical trials database (from compliance, exposure rates and QoL modules of the CRF), but concomitant medication use might be taken from published sources. Collecting concomitant data in a CRF can be an arduous task in a clinical trial because the data often need to be queried for items such as missing dates, details of mode of administration, converting names from active ingredient to either brand or generic name (brands are more expensive) and so forth. In practice, the percentage of patients who take a specific form of concomitant medication might be estimated from the literature rather than using data collected in the trial.

Patient-level safety data will also be collected (e.g. specific adverse event grades for each patient), for which the average cost of treating each adverse event can be extracted from external published sources using reported duration of adverse events. The duration of adverse events gives an idea of *actual* resource use to treat adverse events in the trial. For example, rash is a well-known side effect of Erlotinib (Lee, 2012), treated with an ointment; the incidence of rash can be collected, but details of the treatment for rash can include information such as frequency, dose, name and route of administration, which may not be collected in the CRF. If the time to rash disappearance could be determined from the CRF, this might provide a useful estimate of the duration of concomitant medication use. However, unless the duration of an adverse event is critical, one should restrict computation of duration to the most important adverse events (usually the most costly) to minimise database and programming time (since for duration, the start and stop times of adverse events will be necessary to compute duration, and these are not always available). Lewis (2010) estimates the costs associated with the duration of an adverse event by using concomitant treatment from published data (perhaps because the patient-level data could not be extracted so easily from the clinical trial database). Data managers and clinical programmers would need to ensure that important data such as dates were available for this type of calculation.

In some situations, clinical effects may need to be extrapolated from the clinical trial, even where effects are measured prospectively. For example, 5 years of follow-up time may not be long enough to monitor PFS (progression might occur 6 or 7 years after the start of treatment) in some cancer trials. The mean survival time is often a key statistic required to derive quality-adjusted life years (QALY) for an economic evaluation in cancer. However, this might require knowledge of PFS beyond 5 years. Estimates of PFS times in the specific patient population could be made at Year 6 and beyond based on literature data. Alternatively, as we shall see (Chapter 10), modelling and simulation might be used to estimate PFS effects beyond the fifth year of follow-up; or the follow-up period could just be extended (although this would be more costly) during the design stage of the trial so that extrapolation, and therefore uncertainty, could be minimised. Having observable outcomes is always less uncertain than extrapolating data, whatever the complexity or ingenuity of the models used. If follow-up time is 5 years, but we wish to estimate PFS rates between 6 and 12 years, then modelling and simulation is one approach, and modifying the design of the clinical trial is the other.

Economic evaluation is usually carried out using one (or more) of several methods detailed in Chapter 6. Data summarised from clinical trials are used as inputs into the models to estimate ICERs. For example, in the cost-effectiveness model for Targin, an opioid (Dunlop 2012), opioid-induced constipation (OIC) rates were estimated from the clinical trial based on stochastic analysis, as were estimates of QoL and drug use. However, the estimates of costs (resource use) were obtained from published sources. For the model in Dunlop (2012), costs (external data source), OIC (data from clinical trial), QoL (data from clinical trial) and drug use (data from clinical trial) were used as inputs for the economic evaluation of Targin to estimate the ICER.

3.2.3 Clinical Effects and Outcomes Are Collected Retrospectively

If no clinical trial data are available (data collected prospectively), data are likely to be extracted from published sources. Summaries or even individual patient-level data may be available for analysis. The key difference is that since data collection is retrospective (and a retrospective study can be designed), prospective planning cannot be carried out, and influencing the future (prospective) design of the trial is not possible. One is left to perform an economic evaluation based on whatever data have been collected previously. The quality of the economic evaluation depends on *where* the data were extracted from. In many cases, the retrospective data are reported from published clinical trials such as Dukhovny et al. (2012) or O'Connor et al. (2012). Dukhovny et al. (2012) described their economic evaluation as 'retrospective'; however, the trial was designed prospectively for an economic evaluation. In this book, when we consider retrospective analyses, we take this to mean that the clinical trial did not design for an economic evaluation or collect economic data prospectively, not that the analysis was carried out at some much later time. If the former were true, all analysis (both economic evaluation and statistical analysis) would be retrospective, because analyses are almost always done after the trial has been completed (except perhaps in some sequential and Bayesian-type trial designs, in which analysis is concurrent with trial conduct).

In some situations, prospectively collecting data may be too risky commercially or may just not be the most efficient way. Following an earlier example in OIC, comparing X (experimental) versus S (standard of care), the price requested for the new treatment X may be rejected because it is cheaper to give a laxative (L) combined with treatment S than to pay for treatment X alone. However, if the pharmaceutical company were to undertake a retrospective economic evaluation to show that S + L is not better than S in terms of OIC, a value argument could be presented using retrospective data. Undertaking a prospectively designed trial of S + L versus X might present a commercial risk for the company, as well as being expensive and unnecessary.

One common scenario faced by many pharmaceutical companies is comparing a new treatment A with the current standard of care, treatment S. In a three-arm trial of A, S and Placebo, reimbursement assessors may argue for including S as the third arm to address value for money in terms of superiority (A vs. S). A pharmaceutical company may have its own commercial (or regulatory) reasons to avoid a comparison of A with S (e.g. to minimise impact on market share). One possibility is to carry out A versus S in some indirect way (Ades et al. 2011):

1. Treatment A (the new treatment).
2. Placebo (needed to demonstrate efficacy).
3. Treatment S (needed to demonstrate efficiency and effectiveness). Treatment S has been compared with P in a separate trial.

In this example, consider that treatment S is the competitor drug, and a comparison with treatment A results in A being on average worse than S

(A < S) but not (statistically) inferior to S. It may be commercially risky to compare A with S. An indirect comparison would allow an estimate of treatment benefit between A and S which could otherwise not be made or even known. The so-called mixed treatment comparison enables direct comparisons to be avoided, albeit in a non-controlled setting. A mixed treatment comparison would use published data from a trial that compares treatment S with treatment C as a way to compare the efficiency of A with S. Mixed treatment comparison methods involve complex methods (Chapter 9).

3.3 Clinical Trial Design Issues in an Economic Evaluation

The merits of the RCT over non-RCT designs (e.g. observational-type studies) have been discussed in Drummond (2005). That the RCT evidence needs to be synthesised with all other evidence to arrive at a more informed decision regarding a new treatment's cost-effectiveness (Sculpher, 2006) is still an area of research. For example, Blommestein (2015) describes how an economic evaluation can be accomplished using registry data. There are, however, several points to be taken into account when considering whether a randomised clinical trial is the best design for answering questions of efficiency. Some of these concerns are

- Is an RCT necessary? Not every question requires an RCT to answer it. For example, an observational study might suffice for rare outcomes.
- Is there a consensus on the number and best comparators? If, for example, there are four comparators, this can be an expensive clinical trial.
- Is randomisation acceptable? In some trials, such as surgery or short life expectancy, patients may find it unacceptable to be randomised.
- Adopt more realistic study designs in which usual clinical practice is followed as much as possible.
- Expand the inclusion/exclusion criteria of the clinical trial.
- Avoid the use of placebo unless essential (e.g. unless it is necessary to show dose–response) – active comparators should always be considered. Only when the effect sizes are very small with an active comparator, can a placebo be considered for the purposes of an economic evaluation.
- Is the comparison direct or indirect? If it is indirect, the sample size requirements may be smaller (fewer arms); however, aspects of inclusion/exclusion criteria need to be carefully considered for

the indirect comparator. An indirect comparison may be useless if the two populations are so different that the comparison provides a meaningless conclusion: for example, developers of a new treatment for chronic obstructive pulmonary disease (COPD) may not wish to compare directly with a market leader (e.g. Seretide), and hence an indirect comparison is planned. Comparing the effects of Seretide in COPD patients with the effects (e.g. baseline forced expiratory flow rate in 1 minute [FEV_1]) of the new treatment in asthma patients would be an unreliable comparison (because the two patient populations are different, even if the same end point is used).

3.4 Integrating Economic Evaluation in a Clinical Trial: Considerations

Integrating a health economic component into a clinical trial has been discussed previously (Haycox, 1997; Drummond, 2005; Glick et al., 2006; O'Sullivan et al., 2005; Petrou & Gray, 2011; Ramsey, 2015). We now expand further on the major aspects of trial design for economic evaluation, summarised in Table 3.2.

3.4.1 End Points and Outcomes

Those end points that are clearly clinical, those that are clearly economic and those that have potential overlap should be identified as early as possible. The primary end point is often thought to be unambiguously clinical, whereas QoL may be considered 'somewhere in the middle' between an economic and a clinical end point (Figure 3.1). Compliance end points may be relevant to both clinical (e.g. for per protocol populations) and economic evaluations (cost of actual drug use). The type of QoL measure is also an important consideration: should the QoL measure be disease-specific, like the QLQ-C30 for cancer or the Health Assessment Questionnaire (HAQ) in rheumatoid arthritis, or should a generic measure such as the EQ-5D be used? Careful thought needs to be given as to how important secondary end points can be used to add to the value of a new treatment. Demonstrating value for a new treatment should take into account more than just one end point alone. If the end point is rare, such as chondroma of the mandibular condyle (a benign tumour in the head and neck), a clinical trial may not be the best design.

The QALY mentioned earlier (Section 1.2) is not an end point but a composite of two end points (QoL and clinical effect, often not limited to survival time). This metric is considered by some to be a preferred choice to demonstrate the value argument, although it is not accepted universally (e.g. Germany

TABLE 3.2

Summary of Key Clinical Trial Design Issues for a Health Economic Evaluation

Key Issue	Key Considerations
End points and outcomes	Identify which end points are truly clinical, which are economic and which may be both
Choice of comparator	Standard of care or another comparator
	Direct versus indirect comparison
Timing of measurements	Open label extensions
	Duration of follow-up
	Capture later and early costs
Trial design	Three-arm trial or two separate trials (efficacy and cost-effectiveness)
	Adaptive designs may have limited follow-up
CRF design	Identify resources to collect
	Focus on big cost items
Power and sample size	Number of patients required to demonstrate cost-effectiveness
	Power specific subgroups where cost-effectiveness most likely
	VOI
Treatment pathway	Identify costs and benefits in the treatment pathway
Generic entry/patent expiry date	Identify dates for project management and speed of preparing reimbursement argument
Compliance	Compliance may be related to side effects and hence more costs
Subgroups/heterogeneity	Identify subgroups a priori, defined where treatment benefit is more likely and greatest
Early ICER/INMB	An early ICER helps to work out where the new treatment is in the cost-effectiveness plane
Multicentre/multinational trial	Country-specific CEAC
	Multilevel modelling techniques

and some other countries do not consider the QALY as the primary basis for demonstrating value). The value argument should ideally not be dependent on a single outcome, even if it is a key measure. In situations when a primary end point is positive but all secondary end points are negative (worse effects), a decision in support of cost-effectiveness may be difficult compared with the case where the secondary end points are, on average, not worse (no difference).

Some generic QoL measures lack sensitivity to detect treatment benefit, such as the EQ-5D in lung cancer patients (Khan et al., in press). The cancer-specific measure QLQ-C30 and the even more specific LC-14 (for lung cancer symptoms) can have greater sensitivity to detect treatment effects. It can be incorrectly concluded that economic efficiency is absent, when the problem lies in the measure of efficiency being used.

Other outcome measures may be considered as an alternative to the QALY in some cases, such as quality of time spent without symptoms of disease and toxicity (Q-TWiST), a special type of QALY (Gelber et al., 1989)

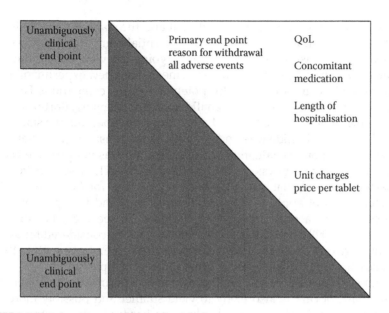

FIGURE 3.1
Relationship between clinical and economic end points in a clinical trial.

that subtracts from the overall survival (OS) the period of time during which either treatment or disease reduces QoL. Disability-adjusted life years (DALYs) and other measures may be considered as outcomes if lack of sensitivity of the measure to detect treatment differences is a possible reason for not demonstrating cost-effectiveness. The failure of an economic evaluation to demonstrate a QALY difference is not sufficient evidence for concluding the absence of an overall value argument. Indeed, in Germany, the Institute for Quality and Efficiency in Healthcare (IQWiG) does not consider QALYs as a prominent measure (or the only measure) that demonstrates the value of the new technology, but any measure that captures a combination of mortality, morbidity and QoL, as shown in the following translated text:

> As the benefit of an intervention should be related to the patient, this assessment is based on the results of studies that have investigated the effects of an intervention on patient-relevant outcomes. In this context, 'patient-relevant' refers to how a patient feels, functions or survives. In this context, consideration is given to both the intentional and unintentional effects of the intervention that allow an assessment of the impact on the following patient-relevant outcomes, in order to determine the changes related to disease and treatment:
>
> 1. Mortality,
> 2. Morbidity (complaints and complications),
> 3. Health-related QoL.... (IQWiG, version 3.0, 27 May 2008)

A comparison between a new treatment and the standard of care is often needed in an economic evaluation. If a potential comparator has not been approved for licensing because it is undergoing a review by the European Medicines Agency (EMEA), this does not mean that a new experimental treatment cannot be compared with this potential (future) comparator. For example, at the time of submission of Lenalidomide for licensing, Bortezamib was undergoing review by the EMEA. Later, Bortezamib became the standard of care while Lenalidomide was under review. The sponsor argued that a comparison for economic evaluation with Bortezamib was not possible because at the time Bortezamib was not the standard of care. The assessors (for reimbursement) argued that although this may be true for licensing, such an argument did not apply to economic evaluations, and the sponsor should have considered a comparison with Bortezamib (see ERG, Lenalidomide submission, 2009). Careful design parameters may be considered for an indirect comparison (mixed treatment comparisons) in situations such as this, so that one can design the trial for a future indirect comparison.

Treatment effect size when comparing a new experimental treatment with the standard of care is also likely to yield smaller and possibly more variable effect sizes. Therefore, it is likely to be harder to demonstrate a cost-effectiveness argument. Comparisons with placebo are often not enough to demonstrate cost-effectiveness, unless either the treatment effect sizes with a comparator are very small or placebo is the standard of care. Toxicity costs for a placebo would not occur in practice and are likely to be artificial costs (O'Sullivan, 2005).

3.4.2 Timing of Measurements

The primary efficacy end point is usually the main focus of a trial, with collection occurring at some 'optimal' set of time points. For other (secondary) end points, such as QoL, the assessment times of collecting data are important. Ensuring that enough pre- and post-disease progression QoL measurements are available will allow more precise QALY estimates; adjusting for pre- and post-progression of disease in cancer trials, for example, would contribute to improved estimates of mean differences in QoL effects. If the median PFS is 6 months, for example, then QoL might be collected every month. However, if the median PFS is only 3 months, more frequent QoL assessments might be made prior to 3 months. For other end points, extended follow-up for monitoring compliance, end of life care, long-term efficacy and safety might also be needed.

In many trials (e.g. cancer trials), QoL is often collected until disease progression, although in theory, it could be collected until death. However, it may be felt that collecting QoL after disease progression is very difficult, because patients will deteriorate rapidly. If some patients continue to take the treatment even after the disease has progressed, time to second progression might be a useful end point. In such cases, having information about QoL between first and second progression will help inform decision makers about the value of treatment over a longer period of time (and justify treatment with the new drug beyond first progression). The period between first and second progression (or death) is essential to estimate the value of a new treatment over the lifetime of patients. Moreover, during the post-progression period, patients tend to take additional concomitant/anti-cancer treatments, which may have a noticeable impact on QoL, because many of these are third- or fourth-line treatments with higher toxicity burdens.

An open label extension after an initial double-blind phase can be helpful to understand real-world effects. During the open label extension, longer-term costs (e.g. for later adverse events, effects of maintenance of new treatment) can be evaluated. Sometimes, after the double-blind phase of a trial, all patients are switched over to the experimental treatment (despite the benefits of the new experimental treatment remaining unproven). Longer-term effects and costs are available for the experimental arm only and not for the comparator.

In some cancer trials, not all patients may be followed till death, so costs are censored for some patients whereas complete costs are available for other patients. Adequate follow-up is needed to capture all costs, later as well as earlier. Follow-up time should not be so long that demonstrating cost-effectiveness and reimbursement is delayed or imprecise. The open label extension should not be glossed over in terms of either design or statistical analysis. The emphasis is often placed on the double-blind part of the trial. The open label part can often be used more efficiently to maximise the value argument, because this part of the trial reflects real clinical practice better than the controlled phase.

3.4.3 Trial Design

Trial design in this context refers to the experimental design of the trial, not the broader aspects of trial design such as blinding, choice of comparator, timing of measurements and so on. Parallel group designs are common in clinical trials. There may be greater efficiency with a three-arm trial (New, Standard and Placebo) than two separate trials, for example New versus Placebo and then another trial of New versus Standard. In some economic evaluations of crossover trials (Grieve et al., 2007), estimating clinical effects takes into account period and carry-over effects through design or analysis. Economic evaluations may also need to take these factors into account when assessing effectiveness. For example, Sharples et al. (2002) observed differences between periods in the economic analysis and carried out 'post-hoc tests of carryover...' for clinical outcomes during an economic evaluation. If costs and effects are correlated, there is no reason why a period or carry-over effect cannot exist when calculating an estimate of the difference in mean costs (i.e. mean differences in costs can also be biased if carry-over or period effects exist).

Novel designs such as adaptive, response adaptive and group sequential-type designs for demonstrating early benefit may limit the possibility of observing medium- to longer-term costs and benefits if the trial stops due to early efficacy. The treatment benefit in such trials would need to be large for the trial to stop early in the first case. However, such designs can tend to focus on statistical significance or posterior probabilities, rather than the mean size of the clinical effect, which is important for economic evaluation. For example, a trial can be stopped early due to a modest clinical effect and smaller than expected variability, which may not be cost-effective. Longer-term measurements for economic evaluations may not be a priority in a sequential/adaptive design, and therefore another separate trial might be needed to demonstrate longer-term benefits/effects, which not only raises ethical issues, but also risks the conclusions of the first trial, especially if it was positive. When trials are stopped early, some additional evidence may be sought from observational-type (real-world evidence) studies or retrospective evaluation of databases (e.g. the Hospital Episodes [HES] data or the National Lung Cancer Audit [NCLA]).

3.4.4 CRF Design

See Section 3.5.

3.4.5 Sample Size and Power

See Chapter 8, where sample size, power and VOI are discussed in detail.

3.4.6 Treatment Pathway

In the same way as clinical pharmacologists attempt to understand the mechanism of action of a drug, health economists seek to understand the mechanism of cost structures. This requires an understanding of the full treatment pathway. Without appreciating the pathway, key component costs could be lost. As an example of a treatment pathway for COPD after a patient has had a confirmed diagnosis and management of his/her COPD has been addressed, the treatment options for stable COPD might be as shown in Figure 3.2.

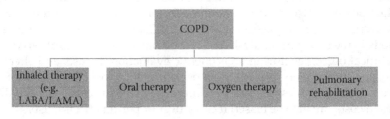

FIGURE 3.2
Example treatment pathway for patients with confirmed COPD.

In Figure 3.2, patients may have several options for treatment, depending on the severity of the condition. In the case of inhaled therapy, for example, short-acting/long-acting beta agonists (SABA/LABA) may be considered, depending on FEV_1. The relative effectiveness of treatments depends on the most appropriate treatment pathway for patients.

3.4.7 Date of Generic Entry

The entry of a generic drug onto the market has huge implications for demonstrating cost-effectiveness. If the reimbursement strategy is not planned adequately, revenue could be lost. Even with 6 months of patent exclusivity for an agreed price of €50 for a 28-day regimen, an estimated market of 20 million patients in Europe can yield in excess of €6 billion sales. For a treatment given approval by EMEA on 1 January 2010 and a reimbursement decision made on 1 December 2010 (because activities relating to health economics were delayed until after approval) and the generic entry date is 1 June 2011, then about 1 year of potential sales revenue is lost. If the cost of the trial is £50 million and revenue in 1 year is £200 million, the impact of preparing reimbursement activities late is not small. This can often happen for in-licensed drugs (treatments bought by a pharmaceutical company, avoiding the costs and resource involved in early phase trials). The relationship between time to licensing, reimbursement and generic entry date is crucial.

3.4.8 Treatment Compliance

Treatment compliance (or lack of) is related to efficacy, safety and efficiency. Lack of compliance may result in lack of effect, or even safety issues, particularly when the therapeutic window is narrow. Patients who take less medication or drop out are likely to be the patients who have more side effects. Non-compliance in these patients is likely to result in greater costs. Patients who drop out also tend to be those who are more ill or in whom there is a lack of effect. Compliance during open label extensions is particularly important to monitor (Hughes et al., 2001), as it may reflect what is likely to happen in real practice. Clinical trials often report trial results using per protocol population analyses, which take into account both per protocol deviations and treatment compliance. Only actual dosing, rather than planned, need be considered, which can contrast sharply with the intention to treat (ITT) definition. In some ITT definitions, efficacy may be based on patients randomised (regardless of whether study medication was taken). Effectiveness should be based on those patients randomised who took study medication.

In some instances, not all the drug is administered, and some is subsequently wasted (sunk cost). For example, some intravenous infusions are very expensive (e.g. if the cost of Rituximab is $500 per infusion). If only 70% of the dose is used in the infusion, the actual cost per milligram used is $350. In economic evaluation, the actual drug taken per patient is used (although one could estimate [or model] future use of treatment over the lifetime of patients, such as in maintenance therapy, if the follow-up period is short).

3.4.9 Identify Subgroups/Heterogeneity

Heterogeneity has been the subject of much interest (ISPOR, 2011; Klaxton). It is often found that some (pre-specified) subgroups of patients in a clinical trial demonstrate greater benefit than the whole sample studied. If a subgroup of patients show a large treatment effect, the argument for cost-effectiveness can be more forceful in this subgroup. For example, Lee et al. (2012) show benefits with Erlotinib compared with the standard of care in those lung cancer patients who showed evidence of rash within the first 28 days. There was no improvement in OS, while there was a modest improvement in PFS, in the whole sample. An economic evaluation (Khan et al., 2015) demonstrated a stronger cost-effectiveness/value argument in this subgroup of patients. Krief et al. (2012) conclude that propensity matching and inverse probability weighting (Chapter 10) are acceptable methods for subgroup cost-effectiveness analysis.

Reimbursement agencies may specifically look for effects in subgroups of patients. For example, in the comparison of Lenalidomide with Dexamethasone, several subgroups were considered; ICERs were presented for each of these subgroups (ERG, 2008). It is not uncommon to see a submission to reimbursement agencies in which various subgroup analyses have been presented. Addressing the heterogeneity of treatment benefit is therefore important not only for reimbursement requirements, but also to identify those subsets of patients who have a higher chance of receiving benefit (and showing cost-effectiveness) with a view to obtaining a premium price for these groups of patients.

The sources of heterogeneity might present through diagnostic testing, pre-specified genetic subgroups (Klaxton, 2011) or even differences in trial results. However, incremental net monetary benefit (INMB) (or incremental health benefit [IHB]) is unlikely to be >0 for every patient within a subgroup. The health benefits and effects are presented as means; therefore, INMB will be either higher or lower than the mean for some patients in that subgroup. At best, we can say that the chance of an INMB >0 is higher in the subgroup of patients compared with patients not in the subgroup. Subgroups should be predefined to avoid data-driven statistical analysis or any other data-driven economic modelling.

Not all heterogeneity may be relevant (Welton, 2011). For example, costs may be heterogeneous between countries, but if decisions on cost-effectiveness need to be made in a specific country, heterogeneity (differences in costs between countries) can be ignored. Heterogeneity in terms of different studies can be minimised if a new study is initiated. For example, if there are 10 trials, each of which estimates an odds ratio as a measure of treatment benefit, then a new trial (the 11th trial) is expected to reduce some uncertainty in the estimate of the overall treatment benefit (because we have more data). The VOI approach compares changes in expected net benefit as a result of adding more data to the existing body of evidence. In this sense, heterogeneity does not always refer to subgroup effects or baseline characteristics that influence the INMB, but, rather, more information.

3.4.10 Early ICER/INMB

There may be interest in generating an early estimate of the ICER/INMB to determine where the new treatment might fall in the cost-effectiveness plane, possibly using Phase II trial data. However, not all Phase II trials collect resource data for cost-effectiveness purposes, although critical safety data and secondary end-point data could be used from Phase II trial data to provide an initial estimate of the ICER. Costs and resource use in a Phase III trial will be different from those in Phase II trials, however. For example, dosing may change, some end points may be dropped or the target population to treat may change. The open label extension of a Phase II trial could, however, provide some insight into how patients respond to the new treatment in real-world situations. Often in Phase II trials, the selected dose is also the same dose used in a larger Phase III trial; in some cases, it is hard to differentiate a Phase II trial and a Phase III trial, other than by sample size. Mitigating risk might involve considering alternative pricing in order to move the treatment out from one quadrant of the cost-effectiveness plane into another by varying parameters such as costs and effects.

3.4.11 Multicentre Trials

One of the realities of clinical trials, particularly large Phase III trials, is that it is very rare to find all patients in a single centre (or sometimes in a single country). For Phase I and small Phase II trials, this might happen, but not in a Phase III trial. Local reimbursement agencies may ask whether patients from their country were represented in the trial; and may raise an eyebrow if in a 10,000 patient multinational Phase III trial, patients from their country were absent. However, for some European countries, it is assumed that treatment effects are generalisable (Willke et al., 1998).

Although generalisability might be true for clinical effects, because the biology and mechanism of action of a treatment should be the same in a homogeneous group of patients whether from Italy or Spain, this may not be true for costs and other outcomes in an economic evaluation. Differences between Italian or Spanish costs of general practitioner (GP) visits, for example, may just reflect underlying policy and societal differences between the two countries. What is of particular interest is whether cost differences between treatments differ between centres (treatment by centre interaction). Such interactions are not powered, and there is some concern that cost differences between centres are not being addressed properly in multicentre trials (Manca et al., 2005; Thompson, 2006; Grieve, 2007).

The specific question of an economic analysis in a multicentre trial is how the INMB varies between centres and how best to capture and then interpret this variability. One approach is to address this concern through modelling techniques such as multilevel modelling. The benefit of using a multilevel (or random effects) model is that an estimate of INMB can be generated for each centre or country. While this may appear attractive and useful in some cases, several considerations should be noted before a multilevel or random effects approach is considered for multicentre analysis in an economic evaluation:

(i) First, it is assumed that centres are sampled at random and are representative of a population of centres. In practice, centres are selected based on competitive recruitment and for other logistical reasons; therefore, centres in a clinical trial are unlikely to be random. In some disease areas, we can be almost certain that centres were not selected randomly (e.g. only a handful of specialist centres may offer stereotactic whole brain radiotherapy).

(ii) Second, although modelling does allow an estimate of the INMB across all centres and addresses the question of whether effects are generalisable across centres, the practical benefits of presenting cost-effectiveness results for each centre for reimbursement purposes can be limited (as well as restricting use if only one subgroup is shown to benefit). This is because pricing is unlikely to be based on different locations within the same country. For example, the UK British National Formulary (BNF) reports pricing for the United Kingdom in general and not by location. It would be awkward to attempt to justify a price that is double or half the price in one region of the country compared with another, especially when sample sizes are limited. However, reporting the cost-effectiveness acceptability curves (CEACs) in a multinational trial by taking into account country-specific costs and effects may have some benefit for local reimbursement decision-making. Careful design considerations, such as the number of patients within each country and the demographic characteristics of each country, should be examined so that reasonable values of the INMB and its precision can be estimated.

(iii) Third, a multilevel model (MLM) or random effects model may not always be necessary. A fixed effects model with a treatment by centre interaction will provide a simple conclusion as to whether the INMB differs between centres. Given the issues raised in (i), generalisability may be less of a concern, because the centres in themselves are seldom randomly selected in a clinical trial. Although there is a loss of (statistical) degrees of freedom if there are many centres, a loss of power to determine the statistical significance of an interaction may not be of great concern, because clinical trials rarely power for cost-effectiveness, let alone interactions for showing cost-effectiveness.

The validity of any assumptions of MLMs need to be critically evaluated before one can be sure about the conclusions (Grieve et al., 2007).

For example, it is assumed that the parameters (e.g. mean costs) are exchangeable (Spiegelhalter, 2003), that is, it is assumed that there are no a priori reasons why mean costs in one centre are higher than another centre, and further, it is assumed that the variability of costs is constant between centres. Reasons for differences in INMB could be due to differences in demographics between centres or countries. For example, in a clinical trial in which there are more patients with a disease diagnosis in one or two centres than in the others, the INMB is likely to be different between centres.

(iv) MLMs might be valid when the data are clustered or correlated within a centre, such that the usual ordinary least squares (OLS) estimates are biased and inefficient (Manca, 2005). But this can be remedied using approaches such as weighted least squares or other approaches for the OLS estimates to remain valid as long as the other conditions of the Gauss–Markov theorem hold. It has also been suggested that MLMs are more appropriate for determining why resource use (e.g. costs) varies between centres (Grieve et al., 2005).

(v) In some cases, Bayesian methods incorporating weak priors are assumed for parameters estimated from an MLM. If the researcher is unfamiliar with Bayesian applications, the considerable complexity involved in choosing a distribution for parameters (e.g. prior distributions for variance components) may be somewhat daunting. Choosing a weak prior is unlikely to offer any material difference from using non-Bayesian methods in multilevel modelling. When there is a strong belief regarding how the distribution of effects and variance components might behave (e.g. using a gamma prior), it is almost certain that the Bayesian approach will offer different results than the frequentist. It is not always straightforward to specify the form of a priori distribution for a treatment effect in a clinical trial; specification of a priori for an effect within a centre may therefore be more challenging (and even more so for costs). However, when it can be done, the methods can demonstrate superior properties (Thompson et al., 2006).

Figure 3.3 shows an example of clustered data in which the INMB (y-axis) shows considerable heterogeneity between centres (x-axis) and in particular varies for males and females across the 30 centres. One might expect the INMB across centres (hospitals where patients are recruited) to be similar (since there may be no real reason why a new treatment is more or less cost-effective in one centre compared with another). In addition, for some sites, the INMB was markedly different between males and females – this might be expected, or, again, might have occurred by chance. A statistical test for differences in INMB between sites, and also a centre by gender interaction, might help address these concerns. One objective of MLM might therefore

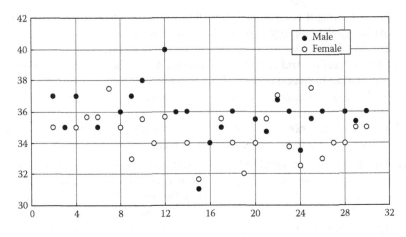

FIGURE 3.3
Clustered data within each centre.

be to adjust observed differences in NMB between treatments for the differences between centres and gender effects.

Example 3.1: Estimating the INMB in a Multicentre Trial in SAS Using an MLM

In this example, we show how to estimate the INMB from observed costs and effects data collected at the individual patient level across three different centres.

FIXED EFFECTS MODEL WITH TREATMENT AND CENTRE AS FIXED EFFECTS

For a fixed effects model with two treatments, the objective is to find an estimate of the INMB after taking into account centre effects. We will estimate the parameters using a fixed effects model with Centre and Treatment fitted as fixed effects. The general representation of the statistical model for this might be

$$\text{Response (NMB)} = \text{Intercept} + \text{Treatment} + \text{Centre} + \text{Error}$$

or, more formally, $Y_{ijk} = \mu + \tau_i + \beta_j + \varepsilon_{ijk}$
where:

Y_{ijk} is the NMB for the ith treatment in the jth centre for the kth patient

τ_i is the 'treatment effect' (for two treatments, this would correspond to the INMB)

(note: since the response is NMB, the word 'effect' refers to differences in terms of NMB and is not the same as a 'treatment effect' in the same sense as a clinical end point)

β_j is the effect of the treatment in the jth centre

ε_{ijk} is the random residual term, independent and identically distributed N $(0, \sigma^2)$.

Since the response is NMB, the word 'effect' refers to differences in NMB (and not the treatment effect in the sense of a clinical end point); the ε_{ij} are independent and identically distributed N $(0, \sigma^2)$.

The SAS code to fit the fixed effects model to the data after first deriving the NMB (i.e. we assume for each patient a cost C_i and a measure of effect E_i to estimate $NMB_i = C_i - \lambda \times E_i$), using the same method as shown in Example 2.3, is

Patient	Treatment	Cost	QALY	Centre	NMB
1	B	720	2.1	1	Use relationship[a]
2	B	1940	2.4	1	
3	B	820	3.1	1	
4	A	250	6.4	1	
5	A	280	7.2	1	
6	B	2540	1.4	1	
7	B	720	1.1	1	
8	B	2940	2.3	1	
9	B	6820	2.8	1	
10	A	650	5.4	1	
11	A	480	6.2	1	
12	B	9540	2.4	1	
13	A	290	7.3	2	
14	A	190	6.5	2	
15	A	280	5.2	2	
16	A	200	8.9	2	
17	B	9290	1.4	2	
18	B	5420	0.9	2	
19	A	290	6.3	2	
20	A	190	5.5	2	
21	A	180	4.2	2	
22	A	100	3.9	2	
23	B	7290	1.4	2	
24	B	6420	0.9	2	
25	A	100	2.9	3	
26	B	290	1.4	3	
27	B	120	0.9	3	
28	A	125	1.9	3	
29	A	990	1.4	3	
30	B	220	0.9	3	
31	A	100	4.9	3	
32	B	190	1.4	3	
33	B	120	0.9	3	
34	A	325	3.9	3	
35	A	890	0.4	3	
36	B	120	0.9	3	

[a] $NMB_i = C_i - \lambda \times E_i$, where E_i are per patient QALYs (or other measure of effect).

```
proc mixed data = x;
class trt centre;
*model NMB= trt centre;
model NMB= trt centre trt*centre / solution;
lsmeans trt centre trt*centre;
run;
```

Output:

Centre	NMB	Mean (£)	SD	Min	Max
1	A	188585.00	22341.78	161350.00	215720.00
	B	62745.00	19309.11	32280.00	92180.00
2	A	179035.00	49364.23	116900.00	266800.00
	B	27395.00	7348.85	20580.00	34710.00
3	A	76578.33	50197.30	11110.00	146900.00
	B	31823.33	7697.07	26780.00	41810

Type 3 Tests of Fixed Effects				
Effect	Num DF	Den DF	F Value	Pr > F
Treatment[a]	1	30	84.02	<.0001
Centre[b]	2	30	13.39	<.0001
Treatment*Centre[c]	2	30	7.82	.0018

[a] Interpreted as showing statistically significant mean differences in the NMB between treatments (i.e. the INMB is significantly higher in one group compared with the other).

[b] There are statistically significant mean differences in the NMB between centres.

[c] The INMB is different between centres.

The first step was to determine whether the NMB (note use of NMB and INMB, the latter refers to the *differences*) differed between centres. From the output, it is clear that there are differences in the mean NMB between centres, but also the INMB differs from centre to centre (Treatment*Centre interaction p-value is <.001). In this situation, the INMB across all centres should not be used. In general, in the presence of an interaction, main effects should be interpreted with caution. The separate question of whether the interaction is qualitative or quantitative is a further issue, for which we need to consider the direction (NMB for A > B or vice versa) of INMBs across centres. However, depending on the nature of the interaction, an estimate of the INMB may still be possible using model-based estimates. If there is no statistically significant Treatment*Centre interaction, then it can be concluded that the mean INMB is similar (not different) across sites (or countries).

MIXED EFFECTS USING TREATMENT AS FIXED AND CENTRE AS A RANDOM EFFECT

In this model, a common mathematical form is presented as

$$Y_{ijk} = \mu + \tau_j + \alpha_k + \varepsilon_{ijk} \qquad (3.1)$$

For each patient (i) on each treatment (j) in each centre (k), the NMB depends on two fixed components: the treatment effect (τ) and the intercept (μ), and two random components: α (centre) with its variance $\sigma^2\alpha$ and the residual effect ε_{ijk} with its variability $\sigma\varepsilon^2$. However we now assume $\varepsilon_{ijk} \sim N (0, \sigma\varepsilon^2)$ and $\alpha_k \sim N(0, \sigma^2\alpha)$. That is, the additional assumption of α_k implies that centres are assumed to be randomly selected from a population of centres.

Yet another way of writing this model (more familiar to multilevel modelling, but equivalent) is as follows: starting from (3.1)

$$Y_{ijk} = \mu + \tau_j + \alpha_k + \varepsilon_{ijk} \tag{3.1}$$

Let $\mu = \gamma_{00}$, $\tau = \gamma_{01}$ and $\alpha = \omega_{0j}$, the model can be written as

$$Y_{ijk} = (\gamma_{00} + \gamma_{01}) + \omega_{0j} + \varepsilon_{ij}$$

The variability of the random components ω_{0j} and ε_{ij} is α_{00} and $\sigma\varepsilon^2$, respectively.

The SAS code for fitting an MLM for these data is

```
proc mixed data = <dataset> covtest;
class centre;
model NMB = treatment / solution ddfm=bw;
random intercept /subject=centre;
run;
```

The term *covtest* gives estimates of the variability of the random effects. The value of $\text{var}(\omega_{0j}) = \alpha_{00} = 1.19$ (i.e. $\sigma\alpha^2$) and $\text{var}(\varepsilon_{ij}) = \sigma\varepsilon^2 = 1.62$.

Output:

Solution for Fixed Effects					
Effect	Estimate	Standard Error	DF	t Value	P-value
Intercept	43476	22197	2	1.96	.1892
Treatment	104635	13916	2	7.52	.0172

Covariance Parameter Estimates (i.e. variance components)	
Cov Parm	Estimate
Centre (σ_α^2)	1.19739
Residual (σ_ε^2)	1.62689

Solution for Random Effects					
Effect	Centre	Estimate	Std Err Pred	DF	t Value	Pr > |t|
Intercept	1	23658	22014	34	1.07	.2901
Intercept	2	13704	22014	34	0.62	.5378
Intercept	3	−37362	21915	34	−1.70	.0973

The fixed effect referring to 'Treatment' provides the INMB value at £104,635. The variance components provide the intra-class correlation coefficient (ICC) = (1.19/1.19 + 1.62) × 100 = 42%, suggesting that centres differ substantially in terms of the NMBs. Confidence intervals can be obtained for the variance components (by adding the option 'cl' in the SAS code). One could also fit the same model separately to QALYs and costs to identify the reasons for differences within and between centres.

The NMB for a particular centre is calculated using the 'Solutions for Random Effects' part of the output. For Patient 4 (on Treatment A) in centre 1, for example, the INMB is estimated as

$$NMB(4) = £43,476 + £104,635 \times Treatment + £23,658$$

The term 4 in NMB(4) refers to the estimated (predicted) NMB for a particular patient (patient number 4) in centre 2 who had Treatment A. For an observed NMB of £191,750, the predicted NMB is £43,476 + £104,635 + £23,658 = £171,769 for Patient 4. For each patient, the estimated NMB within the centre could be calculated and then averaged for each treatment to determine the INMB (by calculating the average for each treatment and computing the difference). The standard errors of the NMB (or INMB) can be used to perform sensitivity analyses for constructing the CEAC.

3.5 CRF Design and Data Management Issues

In clinical trials, data are recorded on a CRF or a DCF. Until relatively recently, CRFs were designed without considering resource collection. Resource use was estimated from surveys or based on external publications. In a clinical trial designed prospectively for economic evaluation, resource data are captured through interviews of patients by nurses or trial staff, sometimes during scheduled clinic visits. Questions such as 'When was the last time you visited the GP since your last visit to the clinic?' or 'Was an outpatient visit to the hospital made?' are included in the CRF. The responses to such questions are then used to calculate the costs per patient. Resource use, not its monetary value, is entered into the CRF. The unit prices of each health resource are obtained from elsewhere (Chapter 4).

Important considerations in CRF design are based on whether micro costing (i.e. all small item costs) or the major costs are to be captured. One approach might be to collect large cost items and assume that the smaller costs are roughly similar between treatment groups. Large cost items might be (for example) hospitalisation, drug use, long-term intensive care stay and concomitant medications. Whenever possible, the common costs should be collected in an unbiased way between treatment groups. Capturing costs more frequently in one treatment group than the other will only cast suspicion on the results. One approach might be to determine which health resource use is likely to make up a large portion of any observed differences in mean costs between the treatment groups before the trial. Reliable estimates of hospitalisation rates will be needed if hospitalisation is the major cost contributor. It may be worthwhile spending time with experts in the varying centres to ask their opinions on where and when costs are likely to occur.

Recording data in a diary may be more suitable for costs, because usually the total cost per patient is derived. Whether these are captured in separate visit pages or only on a single page is not important in practice. Resource use cannot always be planned, unlike scheduled clinic visits. Often a patient might attend a protocol clinic visit, and the question in the CRF might be: 'Since your last visit, did you see a GP?' in order to capture costs associated with GP visits. These unplanned visits could be captured in a patient diary; however, diary data are notorious for being incomplete, unless electronic diaries are used.

Recording data in the CRF or electronic CRF (eCRF) involves patients being asked directly about resource use that has taken place. Sensible judgement is needed to determine whether a diary is needed or whether CRFs can be used to record resource use. It might be useful to run a test or pilot of the CRFs, especially if electronic data capture such as eCRFs is used. Figure 3.4 shows a simple example of a CRF that collects health resource data (Backhouse, 2000).

3.6 Case Study of a Lung Cancer Trial with an Economic Evaluation

We discuss several design issues (leaving analysis for later) of one clinical trial from Lee et al. (2012) as an example. This trial was a Phase III RCT in poor-prognosis patients with non-small cell lung cancer. We discuss and contextualise the important design components involved in this trial. The published cost-effectiveness analysis is found in Khan et al. (2015).

3.6.1 Trial Design

This study was a double-blind, placebo-controlled parallel group study with two treatment groups: Erlotinib (experimental) with Best Supportive Care/ Placebo (BSC) in patients with non-small cell lung cancer. Several interim analyses were planned. The results of interim analyses did not influence the way the final economic evaluation was performed.

3.6.2 Treatment Pathway

These patients were older (>70 years), poor-prognosis patients who were not fit for chemotherapy as they were very frail. If patients progressed on

Since the last protocol visit, has the subject utilized any of the following resources			
Inpatient medical care	☐ 0 – No	1 – Yes	(Complete Hospitalization form)
Outpatient medical care	☐ 0 – No	1 – Yes	(Complete Outpatient Medical Care form)
Outpatient counselling	☐ 0 – No	1 – Yes	(Complete Outpatient Counselling form)
Care from nurse	☐ 0 – No	1 – Yes	(Complete Outpatient Medical Care: multiple visit form)
Care from physical therapist	☐ 0 – No	1 – Yes	(Complete Outpatient Medical Care: multiple visit form)
Care from occupational therapist	☐ 0 – No	1 – Yes	(Complete Outpatient Medical Care: multiple visit form)
Care from speech therapist	☐ 0 – No	1 – Yes	(Complete Outpatient Medical Care: multiple visit form)
Care from health visitor	☐ 0 – No	1 – Yes	(Complete Outpatient Medical Care: multiple visit form)
Care from social worker	☐ 0 – No	1 – Yes	(Complete Outpatient Medical Care: multiple visit form)
Care from other health care worker	☐ 0 – No	1 – Yes	(Complete Outpatient Medical Care: multiple visit form)
Any assistive aids	☐ 0 – No	1 – Yes	(Complete Other Paid Home Care form)
Time in non-acute care institution	☐ 0 – No	1 – Yes	(Complete Institutionalization form)
Time in day center	☐ 0 – No	1 – Yes	(Complete Outpatient Medical Care: multiple visit form)
Any home modification for medical reasons	☐ 0 – No	1 – Yes	(Complete Other Paid Home Care form)
Formal (paid) home care	☐ 0 – No	1 – Yes	(Complete Other Paid Home Care form)
Non-paid caregiver	☐ 0 – No	1 – Yes	(Complete unpaid Caregiver form)
Since the last protocal visit, was the subject			
Involved in any other research activities	☐ 0 – No	1 – Yes	(Complete Concurrent Research form)
Terminated from the study	☐ 0 – No	1 – Yes	(Complete Study Completion form)
Data collection (any) terminated for this patient	☐ 0 – No	1 – Yes	(Complete Data Collection Discontinuation form)

FIGURE 3.4
Example CRF designs for economic evaluation.

the experimental treatment, they discontinued, and subsequent treatment options were few (often palliative care). At the time of trial design and analysis, Erlotinib was recommended for patients who had a genetic mutation called 'EGFR +ve'.

3.6.3 Choice of Comparator

The comparator (BSC/Placebo) was selected because there were no other treatment options for these patients, since they were unfit for chemotherapy.

3.6.4 End Points and Outcomes

This trial used OS as the primary efficacy end point. PFS was a key secondary end point. QoL was captured at important time points before and after expected progression. We estimated that median PFS was about 3 months, and therefore ensured that monthly QoL assessments of EQ-5D-3L (generic) and EORTC-QLQ-C30 (a cancer-specific measure of QoL) were captured. In retrospect, it might have been useful to capture EQ-5D every 2 weeks due to the short survival times of some patients (a proportion of patients died before the first monthly assessment could be made). A balance had to be set between the scientific and economic value of asking patients to complete QoL questionnaires frequently and the patients not being overburdened with forms to complete.

3.6.5 Power and Sample Size

The sample size was 670 patients, with 350 in the Erlotinib group and 320 in the Placebo group. The sample size was powered to detect a difference in OS between treatments. Sample size based on economic evaluation was not considered. However, several subgroups were defined; in particular, those patients who developed rash in the first cycle were the ones who benefited most. In a subgroup of patients who took Erlotinib and did not get a rash, there was no benefit. The sample sizes for some subgroups were smaller, and it might have been useful to have powered these subgroups (a priori). From a cost-effectiveness perspective, those patients who were unlikely to get a rash were considered to be those for whom costs were potentially higher (due to lack of effect). In addition, this was not a subgroup that could be defined at baseline (rash is observed after randomisation), hence, the randomisation was 'broken' for the subgroup comparison of interest.

3.6.6 Date of Generic Entry

This was not relevant, as this trial was an academically led trial. There were several years remaining before the drug would come off patent. Reimbursement claims for Erlotinib had already been made, and data from this trial could have been used to extend the value argument for this particular patient population.

3.6.7 Follow-Up Time (Time Horizon)

The follow-up time was 1 year. The median OS in this patient population was only 3 months. The observation period was long enough to capture the key costs and effects until death, because most, if not all, patients were expected to have died by 12 months. Individual patient-level costs were therefore calculable over the lifetime of patients in the trial.

3.6.8 Compliance

The treatment was one tablet a day, and the numbers of tablets taken and returned were recorded in the CRF. Compliance could have also been measured by a measure called *dose intensity* (which compares planned with actual dosing and is reported as a percentage). However, with dose intensity, taking into account dose interruptions and dose reductions can be complicated. A third possible method involved computing the time from first dose to last dose (if available). Since Erlotinib was dosed once a day, the period of time between first and last doses might be used to determine drug use. Time till discontinuation is not a useful indicator of compliance in some cases. For example, if a patient took study medication for 5 months and was followed up for an extra 4 months for safety and toxicity, and at 9 months the patient discontinued, the total duration of 9 months does not reflect drug use, but only how long the patient was in the study.

In general, however, the cost per patient is determined by the total number of tablets dispensed minus the total number of tablets returned multiplied by the cost per tablet. Compliance may often be skewed, because patients who die early may have 100% compliance. For example, a patient who dies after 1 week, having taken all doses of Erlotinib, is considered to have better compliance than a patient who was in the study for 6 months with 95% compliance. Since economic evaluations are based on mean inputs, in situations such as these, the mean compliance may not be realistic.

3.6.9 Timing of Measurements

The QoL data were collected at scheduled monthly visits. Adverse events were collected as and when they occurred. Resource use at each (monthly) visit was recorded: the number of times a resource was used (e.g. GP visit for moderate or severe rash) post-randomisation. That is, at each monthly clinic visit, the resource use pages of the CRF were completed. In a sense, this is also retrospective, because patients are asked to recall (retrospectively) what resource they used. In some cases, a daily diary is provided. The nominal time points (i.e. those recorded in the CRF as 'month 1', 'month 2' etc.) were not used for the analysis, because the actual times (months in which resource use occurred) were derived for the cost-effectiveness analysis. The follow-up period was long enough to reflect real clinical practice and capture relevant costs and benefits.

3.6.10 CRF Design

The CRF page used in the trial, shown in Figure 3.5, has some clear limitations. Limited patient-level resource use data were collected. At the very least, a tick box identifying major resource use should have been included. In a more extreme case, all details of the potential costs could be collected (such

▲ ADDITIONAL ANTI-CANCER THERAPY

Has the patient received any additional anti-cancer treatment since last assessment? YES ☐ NO ☐

▲ RESOURCE USE SINCE LAST ASSESSMENT (please enter '0' if none) - INCLUDE THIS VISIT

How many nights has the patient spent in hospital?

Medical General/Acute ☐ Date of admission __/__/____ (dd/mm/yyyy)

Surgical General/Acute ☐ Date of admission __/__/____ (dd/mm/yyyy)

Hospice/Respite ☐ Date of admission __/__/____ (dd/mm/yyyy)

HDU/ICU ☐ Date of admission __/__/____ (dd/mm/yyyy)

How many days has the patient attended hospital as an: Outpatient ☐ Day Case ☐

Form Completed by.. Date __/__/____ (dd/mm/yyyy)

Office use –

Form received - date .... Data entered - date .... -initials .... Checked-
date .... -initials ....

FIGURE 3.5
Example CRF from a cancer trial.

as date and duration of visit to the hospital). If the cost per hospital visit is by the hour, then duration measured in days may lead to overestimation of costs; in most cases, the cost per visit and not the duration of the visit is recorded. For example, the cost of GP time might be valued at £80 per visit, regardless of whether this takes 1 hour or 15 minutes. On the other hand, planning of radiotherapy can take several visits, and costs would be cumulative.

3.6.11 Identify and Understand Treatment Pathway

See Figure 3.6.

3.6.12 Early ICER/INMB

This was not relevant in this trial, because it was an academic trial, and pricing was not important. If this had been a commercial trial, early estimates of the ICER/INMB (e.g. using the interim data) could have been used for demonstrating value within the rash subgroup.

3.6.13 Define Subgroups

Several subgroups were predefined. One key subgroup was rash. The cost-effectiveness hypothesis proposed treating those patients who develop rash within the first 28 days of treatment. If no rash occurs with the first treatment

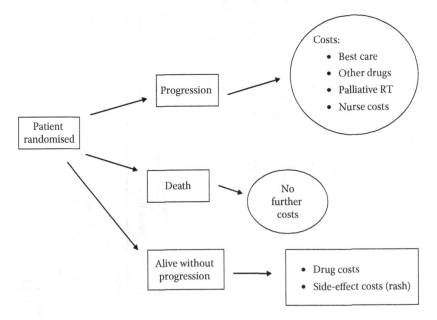

FIGURE 3.6
Treatment pathway (costs) for TOPICAL trial.

cycle, then treatment should be stopped. The value of Erlotinib compared with BSC/Placebo is better demonstrated in patients who develop rash after being treated with Erlotinib for the first 28 days. Costs of treatment for rash, as well as duration of treatment, are important, because median OS and PFS will be higher in the patients with rash. Rash as a biomarker removes the need for a biopsy, which might improve QoL.

3.6.14 Multinational and Multicentre Trials

This trial was a UK multicentre trial, and costs between centres were compared for both treatment groups. Costs were modelled using methods described in Chapters 4, 7 and 10. Because rash was a subjective grading assessment, there were differences between centres in terms of rash severity likely due to differences in opinions between investigators. The greater the severity of the rash, the higher the costs of treating it.

Exercises for Chapter 3

1. Is an economic evaluation retrospective or prospective? Discuss.

2. How would you go about designing an economic evaluation for a clinical trial? What are the important points you would note when the clinical study protocol or grant application is being written?

3. Investigate the treatment pathway for a disease area of your choice and plan an economic evaluation that will be submitted to a local reimbursement authority. What factors would you consider?

4. Why is a mixed effects model used for estimating mean costs from a multinational clinical trial? How would you handle the costs and effects in an economic evaluation from a clinical trial that has been conducted in several countries?

5. Design a CRF for an economic evaluation of a new treatment for a rare disease. What problems might you encounter when performing an economic evaluation?

Appendix 3A: SAS/STATA Code

3A.1 Modelling Centre Effects with SAS

SAS code for modelling NMB for centre effects (or country) interactions.

```
proc mixed data = x;
class trt centre;
*model NMB= trt centre;
model NMB= trt centre trt*centre / solution;
run;
```

3A.2 Modelling Centre Effects as Random with SAS

```
proc mixed data = x covtest;
class centre;
model NMB = treatment / solution ddfm=bw;
random intercept /subject=centre;
run;
```

3A.3 Modelling Centre Effects with STATA

The STATA command for an interaction after coding centres as dummy variables is

```
. xi:regress NMB i.treatment*centre
```

For a mixed effects (random effects), the **xtmixed** command is used:

```
. xtmixed treatment || centre fits the model: NMBij = βXij + εij
```
$+ u_i + v_{j(i)}$ where the $\mathbf{X}i_j$ are the fixed effects for treatment and the $\mathbf{u}_i$ the random effects for centre with ε_{ij} and $v_{j(i)}$ the residual and random error terms respectively.

4

Analysing Cost Data Collected in a Clinical Trial

In this chapter, the different types of costs accrued in a clinical trial are identified. We highlight some of the issues involved in adjusting costs, such as discounting, and then discuss the distributions of costs (including transformation of cost data), followed by an evaluation of how costs can be modelled using the more common and some complex approaches. We finally consider some practical issues in handling missing costs. A table is also provided which identifies where health resource costs might be available for several countries.

4.1 Collecting and Measuring Costs for the Case Report Form

4.1.1 Costs versus Resource Use

In clinical trials with a cost-effectiveness objective, an estimate of mean costs is essential to calculate the incremental cost-effectiveness ratio (ICER) or incremental net monetary benefit (INMB). In this chapter, we discuss the issues involved when collecting, measuring and analysing patient-level costs in a clinical trial. However, we first need to understand what is meant by 'cost'. Expertise from a health economist can be valuable, because identifying some healthcare costs requires one to think in economic terms when undertaking an economic evaluation.

One might think of 'costs' in rather simple terms, such as the money paid for certain goods purchased. However, identifying costs in monetary terms for a clinical trial is not easy to conceptualise. For example, it would be hard (and unnecessary) for nurses or site monitors to enter the prices of each laboratory test into a case report form (CRF) or for patients to haggle over the costs of a computed tomography (CT) scan or the standard of service (care) they were receiving. The alternative is to think in terms of how much *resource* has been used. Specific outcomes (or measures), such as hospitalisations, concomitant medications, CT scans, consultation time and amount of drug taken, will often be reported in the safety sections of the clinical study report (CSR). The quantity in milligrams of study drug taken and the number of hospitalisation episodes are the items typically reported in the CSR.

TABLE 4.1

Example of Deriving Patient-Level Costs from Resource Use Data

Patient	Blood Sample[a]	Cost/Sample	Surgery	Cost of Surgeon Time	Total Cost
1	Yes	£50	Yes	£1000	£1050
2	Yes	£50	Yes	£2000	£2050
3	Yes	£25	Yes	£1000	£1025
4	Yes	£50	Yes	£1500	£1550
4	Yes[a]	£50	Yes	0 (as above)	£50
Total cost					**£5725**
Mean cost					**£1431**

[a] A repeat sample was required for this patient (e.g. because of toxicity, in which case it can be considered as a treatment-related cost).

These quantities are subsequently converted to monetary values for use in later economic evaluations. Therefore, it is not costs which are collected but *resource use* which is central to the data collection, management and data-cleaning process in clinical trials.

Once the quantity of resource use has been identified, the next task is to *value* the resource use, which depends on the unit *price*. The cost is the unit price multiplied by the quantity of resource. This will yield the patient-level cost used for later modelling. For example, if the price of a blood test is £25, and three additional tests (four in total) are required for that patient, the total cost of the test will be £100 (Table 4.1 provides examples of sources from which unit prices may be obtained for several countries). The mean costs (expected costs) are then computed across all the patients in the trial. At the patient level, one would sum across the costs for each patient first to derive a per patient cost and then compute the mean costs between patients (Example 4.1). In this chapter, we are primarily interested in how to analyse the patient-level cost data collected in the trial.

Although we may not be primarily interested in modelling resource use, for example when the outcome for analysis is not a monetary value but is measured in terms of the number of hospital visits or the amount of drug taken per patient, econometric models can be used to estimate the mean resource use if patient-level resource use data are available (Deb et al., 2000; Raikou et al., 2004). Consequently, the mean resource use is then multiplied by the unit price so that the mean cost can be determined. This is different from converting all the resource use into monetary values and performing the modelling on the costs. The two approaches will not yield the same estimate of mean costs.

Example 4.1: Alternative Approaches to Deriving Total Costs

Table 4.1 shows one approach to deriving the total costs from patient-level data, including resource use (blood samples and surgery use) for a given treatment arm. The total costs are first estimated at the patient level and then averaged. The total cost for all patients is £5725 (£1050 + £2050 + £1025 + £1550 + £50). The mean per patient cost associated with treatment is calculated as £5725/4 = £1431. If we calculate the mean costs of the blood samples as £56 (i.e. £50 + £50 + £25 + £100/4) and then add this to the mean costs of surgery (£5500/4 = £1375), the average cost will be the same (£56 + £1375 = £1431). However, if the blood sample cost of Patient 3 was 0, the two approaches of calculating mean costs would differ. This is a naïve approach to determining mean costs, and modelling approaches accounting for other factors such as age, country, repeated visits and so on (Section 4.5) should be used.

In some cost-effectiveness analyses, resource use, such as nurse time and general practitioner (GP) time, may not have been planned prospectively, in which case retrospective resource use is estimated from the published literature. Other resource use, such as drug exposure (number of tablets), can be collected in the trial even if the outcomes are not considered a 'cost outcome'. Common safety outcomes (e.g. adverse events) can also be important cost drivers. By thinking of safety end points as 'cost outcomes' for reimbursement purposes, one may find that there is likely to be a more thorough interrogation of safety data than is otherwise undertaken (often relegated to simple summary statistics); if safety data can be thought of as influencing the cost-effectiveness of the drug, they are likely to be scrutinised in more depth rather than presented as simple summary data. Safety data, from a purely clinical perspective, are more important when large or meaningful (or even statistically significant) differences occur between treatments. This may not be true from a cost-effectiveness perspective, in which even small differences in some safety parameters can influence cost.

For example, if the rate of hospitalisation is 5% versus 7% for two treatments A and B, the difference might be too small to be statistically or even clinically meaningful. If the mean cost associated with Treatments A and B (ignoring any hospitalisations for the moment) is £35,000 and £32,000, respectively (an incremental cost difference of £3,000 in favour of Treatment B), and if the cost of hospitalisation is, on average, £8,000 per patient for Treatment A and £5,000 for Treatment B, then the new mean costs are £43,000 for Treatment A (£35,000 + £8,000) and £37,000 (£32,000 + £5,000) for Treatment B: an incremental cost difference of £6,000 in contrast to the previous £3,000. The change in incremental costs from £3,000 to £6,000 is the result of a 2% difference in hospitalisation rates considered clinically irrelevant. Differences between treatments for some safety end points are often ignored because sample sizes are usually too small to detect statistically significant differences. In cost-effectiveness analyses, the lack of statistical significance may be irrelevant (Claxton, 1999), because observed treatment differences are interpreted in the context of 'value' rather than whether they are statistically different. A treatment difference may be clinically relevant but may not however show statistical significance, in which there may be less concern, especially if the lack of significance is borderline (e.g. a p-value of 0.059). Similarly, a lack of (or the presence of) statistical significance between two values of NMB does not change the conclusion as to whether the INMB and consequent ICER is cost-effective and neither would statistical significance alter the interpretation of whether the new treatment offers value or not. The implications of this will necessitate planning ahead and determining which parts of the safety data (safety tables and listings) can be elaborated or further detailed to meet reimbursement needs when writing the statistical analysis plan (SAP). Perhaps a reimbursement analysis plan (RAP) or a payer analysis plan (PAP) could be developed which identifies statistical analyses required for local reimbursement bodies (before final analysis).

4.2 Types of Costs

The important costs are ultimately those which the payers consider important (and not necessarily the clinician, health economist or statistician). Therefore, trials should be designed such that the payer perspective is taken into consideration when identifying costs. The practical difficulty of this is that in multinational trials with differing local payer perspectives, it can be challenging to meet all payer needs in a single clinical trial. Nevertheless, the two most important cost components which all payers are likely to consider essential to determining value are termed *direct* and *indirect costs*. Direct costs can be further classified into direct medical and direct non-medical costs. Examples of these costs are shown in Figure 4.1.

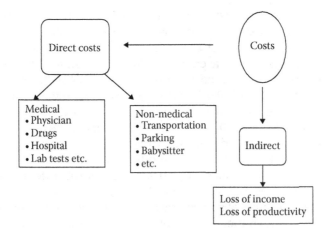

FIGURE 4.1
Examples of direct and indirect costs.

In a clinical trial, the direct costs are items such as study drug use (cost per tablet, per milligram or per infusion), staff costs (e.g. nurse time for administering infusions in the case of intravenous treatment), consultation time, electrocardiographic (ECG) equipment use and ultrasound equipment. In situations where hospitalisation is an important cost, items such as hospital buildings might also be considered, but this may be less relevant for a clinical trial. In most instances, the resource use should reflect what happens in standard clinical practice. In some multicentre clinical trials, some sites may be more efficient than others, for example some private hospitals may have longer (more thorough) consultations with higher unit prices. Another level of complexity may occur when the majority of patients are recruited from one country (or site) and a few from other countries, particularly those countries not considered to be part of the payer perspective.

For example, in a multinational clinical trial of gemcitabine and docetaxel compared with doxorubicin as first-line treatment in previously untreated advanced unresectable or metastatic soft tissue sarcomas (e.g. the Cancer Research UK GEDDIS trial), an economic evaluation was planned during protocol design. The economic evaluation was planned from a UK National Health Service (NHS) payer perspective, with the majority of patients recruited from UK sites. However, due to recruitment problems, a few sites were opened in Switzerland (only UK and Swiss sites recruited patients). The question was whether it was relevant to collect health resource data from the Swiss sites.

On the one hand, for the purposes of efficacy analysis, it was planned to analyse data from all patients. The economic evaluation would therefore use efficacy data from the Swiss patients. On the other hand, for costs analysis, the resource values were likely to be very different between Switzerland and the United Kingdom due to differences in health systems and practices. Therefore, mean costs were based on the UK sample of patients only, and the Swiss data were ignored. Although efficacy data were used across both countries to determine the denominator of the ICER, additional statistical tests (including additional sensitivity analyses) were planned (for 'flagging' purposes) to determine whether the mean costs differed between the UK and Switzerland.

4.3 Other Concepts in Costs: Time Horizon and Discounting

There are several important concepts relating to costs which are important when valuing healthcare resource. It is useful for the researcher to know about these concepts, two of which are briefly highlighted in Sections 4.3.1 and 4.3.2. Readers who wish to know more about these and other concepts should consult Culyer (2011).

4.3.1 Time Horizon

The time horizon is important because it reflects the period of time for which an item of health resource is used or is expected to be used. The period of time does not just reflect what happened during the trial but also what is *likely* to happen beyond the trial. For example, in some clinical trials, an 'end of trial' declaration (a statement which defines when the end of trial occurs, e.g. when the last patient has had his or her last follow-up visit) is included in the protocol. After this time, collection of patient-level clinical trial data may not be possible (due to local regulatory laws). On the one hand, in practice, running a trial with a long follow-up may be costly because of the need to employ staff to collect, clean and report safety and other data. However, not collecting long-term safety, efficacy and quality-of-life data can miss important information which could influence the calculation of mean costs. For practical purposes, follow-up time in a clinical trial for collecting resource use coincides with the length of follow-up for efficacy and safety data, which may not necessarily be the best solution, but is often convenient. Observing resource use for a long period of time in the experimental treatment arm and shorter periods for the control arm will result in biased estimates of incremental costs.

From an economic perspective, the time horizon should be such that resource use is captured during most, if not all, of the follow-up period of a clinical trial *and, where possible, beyond*. For example, in a clinical trial in which patients have an expected survival time of 10 years, follow-up might be restricted to 3 years for practical purposes. After 3 years, estimates of resource use are taken from the published literature or modelled. Even if it is not possible to follow up all patients for 10 years, collecting data on a few patients beyond 3 years (but not necessarily as far as 10 years) can facilitate extrapolated estimates of costs if statistical models are used. For example, if some patients are followed for 5 years and some patients are followed for 7 years, it might allow a statistical model to predict the extra 2 years of costs in those patients who are still alive at 5 years (but are no longer followed up).

Another example might be a clinical trial with a mortality end point, in which the follow-up is planned for 2 years and >95% of patients die within 2 years. In this case, the time horizon can be considered as the period between randomisation (or date of first dose) and death for each patient. Resource use (costs analysis) evaluated over 2 years might be acceptable. However, in a clinical trial of pain in which follow-up is over 52 weeks but the primary end point is at 24 weeks, a suitable time horizon might be 1 year.

It is likely that payers would like to see evidence that the value of a new (experimental) treatment is reflected using a time horizon related to the duration of the disease and not a time horizon which provides evidence of efficacy and safety only over a shorter period of time. One limitation of long follow-up windows is that it can be difficult to know whether costs associated with long-term toxicities are solely due to the drug under investigation or are a result of other drugs taken concomitantly. Collecting concomitant medication data in a clinical trial with a shorter follow-up can be difficult (e.g. due to missing dates and modes of administration), and therefore collecting longer-term data of this nature is likely to be even more complicated.

4.3.2 Discounting

We briefly introduced the concept of discounting in Chapter 1, where future costs were valued in present terms. For example, if surgeries are expected to cost £5,000 per year for the next (future) 5 years, the cost of surgery in 5 years relative to the present time will be less than £5,000. This is not because surgeons will reduce the price of carrying out surgery, but it is due to the concept of *time preference*. For example, some people might prefer to smoke today and not do regular exercise because they place greater value on their leisure time now than later; others might want to keep fit in the hope that future benefits in terms of a healthier and longer life might be obtained. Similarly, there is also a concept of time preference for costs, where the preference is for having a dollar today rather than having a dollar tomorrow. Intuitively, one might expect that the price of surgery will increase in 5 years' time. However, the two concepts are linked, because in 5 years' time, £5,000 may not be enough to pay for the surgery. From the perspective of the surgeon, the relative value of receiving £5,000 for performing a surgery today will be greater than if he or she received £5,000 in 5 years' time. Similarly, the value of receiving $1,000 tomorrow is not considered to be the same as having $1,000 today (or in 5 years' time); tomorrow's (or future value of) $1,000 has been discounted (or devalued). Using a discounting rate of 4% per year, for example, the cost of surgery costing £5,000 now would be valued in 5 years' time at

$$£5,000 / (1 + 0.04)5 = £4,274$$

For the purposes of clinical trials, this concept refers to future costs and benefits. Therefore, we would need estimates of future costs and benefits, often derived from modelling clinical trial data. Discounting is therefore employed when future benefits and costs are expected to occur outside the clinical trial time horizon.

In a situation in which some trials are conducted over a long period of time, such that costs change from the earlier part of the trial to a later part, there may be a need to adjust (inflate) earlier costs. For example, if the analysis is carried out in 2012 and the trial started in 2001, then all costs might be inflated (indexed) to 2012 prices (prices are increased) so that all costs reflect the present value of resource use. This is different from discounting, in which future benefits and costs are adjusted downwards.

4.4 CRFs for Collecting Resource Use Data in Clinical Trials

An important aspect of costs analysis is data collection. In clinical trials, the CRF is a critical document which is reviewed by several professionals to ensure that clinical data can be collected accurately. This is no less true of CRFs designed for resource use. In this section, we give several examples of CRFs which can be used in clinical trials (Figures 4.2 and 4.3). These CRF designs are examples, and modifications can be made to accommodate specific trial needs. One important consideration when collecting resource use data in a clinical trial is unscheduled visits, such as additional GP visits. Not all unscheduled visits are direct treatment costs. For example, a repeat visit for a laboratory test is not necessarily a direct treatment cost. Therefore, a distinction needs to be made between collecting the resource use associated with treating patients and the costs of running the clinical trial. Where adverse events are collected for reporting incidence rates, the treatment-emergent adverse events (TEAEs) are most important, because these again are associated directly with the costs of treatment (because these adverse events emerge after treatment has been taken).

4.5 Statistical Modelling of Cost Data

In this section, we discuss several issues in the statistical modelling of patient-level cost data. We assume that for each patient there is a record of health resource used converted into monetary value (values ≥0). Although statistical models can be used to estimate mean resource use (e.g. the number of GP visits and amount of drug taken), this chapter is concerned with models in which the response (dependent variable) is measured in monetary units. In some cases, the objective might be to estimate, for example, the

▸ RESOURCE USE SINCE LAST ASSESSMENT (please enter '0' if none) - INCLUDE THIS VISIT

How many nights has the patient spent in hospital?

Medical general/acute	☐	Date of admission __/__/____	*(dd/mm/yyyy)*
Surgical general/acute	☐	Date of admission __/__/____	*(dd/mm/yyyy)*
Hospice/respite	☐	Date of admission __/__/____	*(dd/mm/yyyy)*
HDU/ICU	☐	Date of admission __/__/____	*(dd/mm/yyyy)*

How many days has the patient attended hospital as an: Outpatient ☐ Day Case ☐

FIGURE 4.2
Example CRF from a lung cancer trial.

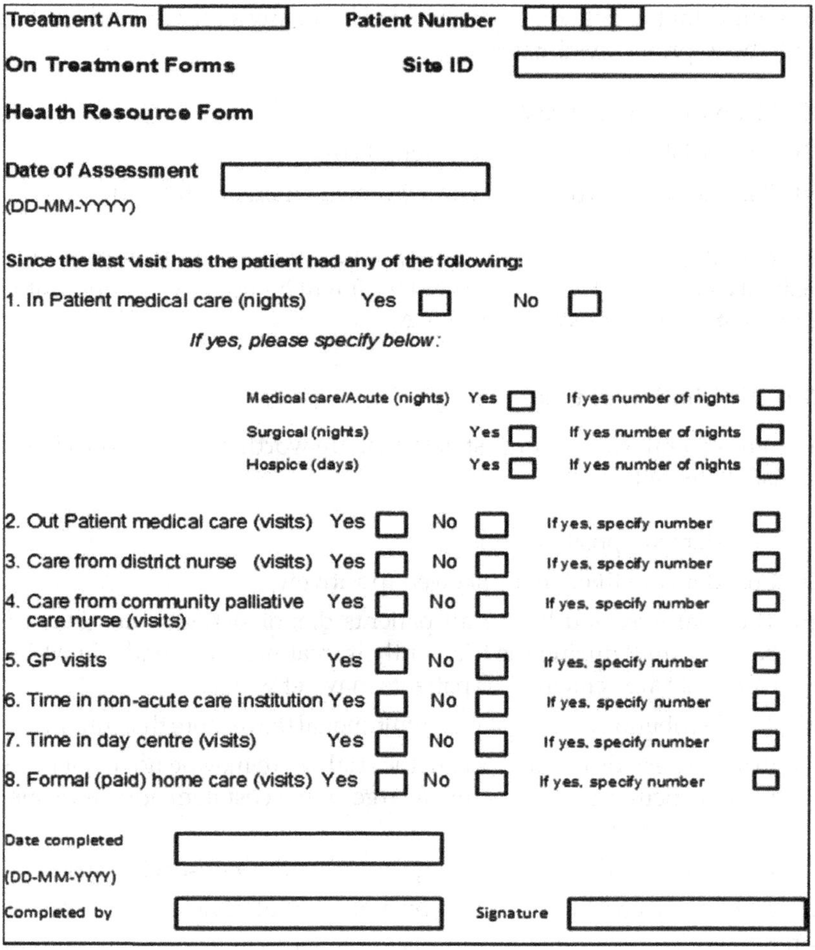

Treatment Arm ☐	Patient Number ☐☐☐☐
On Treatment Forms	Site ID ☐
Health Resource Form	

Date of Assessment ☐
(DD-MM-YYYY)

Since the last visit has the patient had any of the following:

1. In Patient medical care (nights) Yes ☐ No ☐

   If yes, please specify below:

   Medical care/Acute (nights) Yes ☐ If yes number of nights ☐
   Surgical (nights) Yes ☐ If yes number of nights ☐
   Hospice (days) Yes ☐ If yes number of nights ☐

2. Out Patient medical care (visits) Yes ☐ No ☐ If yes, specify number ☐
3. Care from district nurse (visits) Yes ☐ No ☐ If yes, specify number ☐
4. Care from community palliative care nurse (visits) Yes ☐ No ☐ If yes, specify number ☐
5. GP visits Yes ☐ No ☐ If yes, specify number ☐
6. Time in non-acute care institution Yes ☐ No ☐ If yes, specify number ☐
7. Time in day centre (visits) Yes ☐ No ☐ If yes, specify number ☐
8. Formal (paid) home care (visits) Yes ☐ No ☐ If yes, specify number ☐

Date completed ☐
(DD-MM-YYYY)

Completed by ☐ Signature ☐

FIGURE 4.3
Example CRF from a sarcoma cancer trial.

mean *number* of GP visits (and not the mean costs of GP visits). This approach is not uncommon in econometric-type models.

The focus of our approach will be on estimating the mean cost for each treatment using several different statistical models. We start by discussing the simpler ordinary least squares (OLS) regression models, followed by general linear modelling (GLM), and then discuss other approaches to models, including some newer complex methods. When choosing a particular modelling approach, we reiterate Ockham's razor on the principle of parsimony (select the simplest approach first). The reader should also note carefully Box's (1987) statement that 'All models are wrong, but some are more useful'. Finally, we discuss some important aspects involved in handling cost data, such as missing and censored costs.

It is important to note that the key objective of any statistical model chosen should be to provide estimates of

(i) Mean incremental costs
(ii) The variability of mean incremental costs
(iii) The standard error associated with mean incremental costs

Each of (i)–(iii) is important as an input for carrying out later probabilistic sensitivity analyses (PSA). In particular, (i) will be used as an input into the estimate of the mean INMB $= \Delta_e{}^*\lambda + \Delta_c$.

4.5.1 Distribution of Costs

There are several features of cost data that are worth noting before choosing a statistical model:

(i) Cost data are positive (>0).
(ii) Cost data are likely to be skewed (positively).
(iii) The trial may end before all patients die; or patients may become better (cured) during the trial, or their treatment may end – in which case costs are censored; or patients may get worse.
(iv) The distribution of costs may be multimodal (have more than one peak).
(v) In some periods of time during the trial, there may be no major costs for a particular patient, before a large single cost item appears, causing considerable skewness.
(vi) Following (v), there are varying time periods of observing costs.
(vii) The (mean) costs may depend on a number of covariates.
(viii) The sample mean due to skewness in the data may not be a robust or efficient estimate of the population mean.

One important feature of a health economic evaluation is that no matter the skewness or other features of the data (such as non-normality), we are interested in inferences on the mean costs. It has been suggested that where differences in mean costs are smallest (and differences in mean effects are largest), this yields Kaldor–Hicks efficiency criteria. That is, some aspects (but not all) of Pareto optimality are achieved (Glick et al., 2014).

4.5.2 Analysing Costs Assuming the Data Are Normally Distributed

Cost data are, in general, skewed, multimodal and censored (or missing), and therefore the assumption of normality is unlikely to hold true. However, if the sample size is sufficiently large, then even when cost data are not too extremely skewed, two-sample tests such as the t-test and analysis of variance

(ANOVA) can still perform remarkably well. It is interesting to note that for Phase I bioequivalence trials, in which some features of the data are also similar to cost data (positive, skewed and log-normally distributed), sample sizes as small as 12 are considered sufficient to provide legitimate inferences for market authorisation of generics or switches in formulations (EMEA, 2010).

The fact that some sample size tables and software start with sample sizes of at least six (Machin et al., 2009) per group suggests that parametric tests may be acceptable even when sample sizes are quite small. Given that most economic evaluations are performed in Phase III trials where sample sizes are large, tests based on normality assumptions may hold in most situations, and the data would need to be excessively skewed to render the normal theory 'invalid'. Where the primary outcome is a clinical measure, normal theory helps to reduce the sample size and increase the power for rejecting a null hypothesis, if a hypothesis testing approach is considered appropriate. Sample sizes for cost-effectiveness might be more concerned with generating an unbiased estimate of the mean difference than with testing specific hypotheses. This is important to consider for subgroup analyses if cost-effectiveness analyses are carried out within subgroups. What is perhaps more important is that mean *differences* in costs are normally distributed.

Mihaylova et al. (2011) suggest that under reasonable normal assumptions, standard methods such as t-tests, ANOVA and linear regression (e.g. mixed effects models) can be used to provide an estimate of the mean incremental cost. A simple linear model (basically a two-sample t-test) might be

$$\text{Costs} = \text{Intercept} + \text{Treatment}$$

or

$$C_{ij} = \mu + \tau_j + \varepsilon_{ij}$$

where:
- C_{ij} is the cost for patient i on treatment j
- μ is the common intercept
- τ_j is the incremental cost presented on the original scale
- ε_{ij} is the residual error, assumed to be independent and identically distributed with mean 0 and variance σ^2

This model is different from the one presented in Chapter 2 where we first calculated the NMB for each patient. Here, we are interested in computing the mean incremental cost only (i.e. the value of Δ_c).

A simple SAS code for this model without covariates is

```
Proc glm data = <dataset>;
Class treatment;
Model Cost = Treatment;
Run;
```

Additional covariates (e.g. age and centre) can be added, and a mixed effects model framework can also be used:

```
PROC MIXED data = <dataset>;
Class treatment country;
Model Cost = treatment country;
Random patient;
Lsmeans treatment;
Run;
```

When the sample sizes are small and there is moderate to severe skewness present, this approach may not be possible. Where such skewness is present, methods described in Section 4.6 or non-parametric methods (e.g. bootstrap methods) might be considered; some non-parametric approaches (quantile regression, absolute deviation models) may not yield differences in terms of mean costs and may not be useful in practice.

Example 4.2: Modelling Costs with and without Covariates (Normally Distributed)

The following example of 48 patients with patient-level total costs is used to generate the mean incremental costs before and after adjustment for age and country effects (Table 4.2). In this example, only costs are modelled (ignoring effects). The INMB is estimated using a suitable value of λ, the willingness to pay, in order to generate the base case estimate of the INMB. If the correlation between costs and effects is also to be considered, a more complicated bivariate (two response variables) will be required, and this is shown in Example 4.3. The output is shown in Table 4.3.

Interpretation:

- The mean incremental cost is £6 without accounting for covariates, and increases substantially to £180 when accounting for age and country.
- Age is not associated with higher (or lower) mean costs, but country is associated with costs (i.e. there are significant differences in mean costs between countries). The treatment*country interaction p-value is .549, suggesting that on the whole (across all countries), the incremental cost does not differ significantly.
- We should be careful about the interaction terms, because trials are often not powered to detect interactions (for clinical effects, let alone costs), especially for a sample size as small as this. Table 4.4 shows the country-specific mean costs for each treatment. There is high variability in incremental costs, which partly explains the lack of significance; in addition, in the United Kingdom, the costs of Treatment B are much higher than those for Treatment A. Local payers (e.g. the National Institute for Health and Care Excellence [NICE]) may be more interested

TABLE 4.2

Data for Example 4.2

Patient	Treatment	Country	Age	Cost (£)	Patient	Treatment	Country	Age	Cost (£)
1	A	United Kingdom	34	5,268	25	B	Germany	63	8,126
2	A	United Kingdom	26	1,535	26	B	Germany	34	6,535
3	B	United Kingdom	58	8,261	27	A	Germany	30	12,111
4	A	United Kingdom	64	526	28	A	Germany	59	4,236
5	A	United Kingdom	44	3,126	29	B	Germany	54	7,126
6	A	United Kingdom	32	2,671	30	B	Germany	71	7,671
7	B	United Kingdom	66	4,319	31	A	Italy	66	4,151
8	A	United Kingdom	42	2,111	32	A	Italy	34	4,213
9	B	United Kingdom	34	8,881	33	B	Italy	31	5,481
10	B	United Kingdom	68	9,123	34	B	Italy	54	3,923
11	A	France	74	18,146	35	A	Italy	78	22,146
12	A	France	59	5,111	36	A	Italy	34	25,111
13	B	France	34	5,998	37	A	Italy	57	8,199
14	B	France	62	4,129	38	B	Italy	34	12,129
15	A	France	48	3,112	39	A	Italy	32	13,199
16	A	France	61	2,189	40	B	Italy	34	23,489
17	B	France	34	3,777	41	A	Spain	57	13,797
18	B	France	34	6,453	42	B	Spain	58	16,111
19	A	France	29	7,268	43	A	Spain	33	7,658
20	A	France	44	9,121	44	B	Spain	44	9,311
21	B	Germany	34	4,321	45	B	Spain	79	1,321
22	B	Germany	61	4,139	46	B	Spain	35	1,590
23	A	Germany	34	5,912	47	B	Spain	69	12,917
24	A	Germany	34	3,998	48	B	Spain	34	9,654

TABLE 4.3

Summary Output for Example 4.2

Model	LSMean	Incremental Cost	95% CI	p-value
Without covariates	A 7705	6[a]	(−3356, 3367)	.997
	B 7699			
With covariates	A 7844[b]	180[a]	(−3090, 3450)	.912
	B 7664[b]			
Age				.596
Country				.029
Country*treatment				.549

[a] Difference between A and B: Treatment A has higher mean costs.
[b] After adjustment for covariates (age p-value = 0.596; country p-value = 0.029).

TABLE 4.4

Summary Treatment by Country Mean Costs

Country	Treatment A Cost (£)	Treatment B Cost (£)
France	7,491	5,089
Germany	6,564	6,320
Italy	12,837	11,256
Spain	10,728	8,484
United Kingdom	2,540	7,646

in this aspect of incremental cost than the much smaller overall incremental cost. The mean differences in costs between treatments in Germany are smaller (£244).

Example 4.3: Joint Modelling of Costs and Effects

In this example, we use simulated data (for illustration purposes) using the SAS code provided in Appendix 4A to analyse costs and effects jointly. A total of 1000 cost and effectiveness values are simulated from a normal distribution (two measures per subject). We model the costs and effects jointly in terms of treatment and assume an unstructured correlation between costs and effects (i.e. we do not make any assumptions about how the costs and effects are related, other than what is observed). From a joint model, we would get the mean incremental costs and effects simultaneously (with standard errors) in a single model. Since this section is about costs, only the incremental costs are reported and interpreted. A similar interpretation can also be made for incremental effects.

The data format is best put in the 'long' format, whereby each patient has two responses, as shown in Table 4.5. However, if multivariate analysis of variance (MANOVA) is used, the format will be different (with one column for cost and one for outcome). One of the key reasons for using a MANOVA is to use a more powerful test to reject the null hypothesis of no difference. Since hypothesis testing is of less primary concern in economic

TABLE 4.5

Data for Example 4.3

Patient	Treatment	Age	Outcome	Response
1	A	34	Costs	5268.00
1	A	34	Effect	0.65
2	A	26	Costs	1535.00
2	A	26	Effect	0.96
3	B	58	Costs	8261.00
3 etc.	B	58	Effect	1.48

evaluation, a fixed effects model is chosen, and we compare the standard errors of the means (used as inputs for later PSA). A mixed effects model could also be used, and the SAS code is also provided in Appendix 4A.

A modelling approach similar to that discussed in Manca et al. (2007) is chosen, but using a frequentist method (a non-informative prior, as in a Bayesian setting, will result in very similar estimates of mean incremental effects and costs without the unnecessary complications of justifying the choice of a prior). A simplified description of the bivariate cost-effectiveness modelling is

$$\left(\text{Costs, Effects}\right) = f(\mu_c, \mu_e, \sigma^2_c, \sigma^2_e, \Sigma_{ce}) \qquad (4.1)$$

where μ_c and μ_e are the mean costs and effects, respectively, with variances σ^2_c and σ^2_e and correlation Σ_{ce}. Equation 4.1 describes the functional relationship between the population estimates of mean costs and effects explained through a bivariate normal distribution (BVN) with a specific variance covariance matrix Σ_{ce}. For practical purposes, we only need to be concerned with the suitability of the model (using model fit statistics) and then how to interpret the effects. If no correlation between costs and effects is assumed, then two separate models (for costs and effects) to derive the mean incremental cost and incremental effects can be used.

The SAS code for modelling a bivariate fixed effects model for costs and effects for the data in Table 4.5 is

```
PROC MIXED data =<dataset>;
Class Treatment outcome;
Model Response = outcome treatment outcome*treatment;
Repeated outcome/ type=un grp=outcome*treatment sub=pat;
Lsmeans outcome*treat/cl;

Run;
```

Summary output for Example 4.3 is given in Table 4.6.

Interpretation:

After modelling the relationship between costs and effects, without specifying the nature of the correlation between costs and effects (type = UN in the SAS code), both the univariate and the bivariate model

TABLE 4.6

Summary Output for Example 4.3 (Costs Only)

Model	LSMean (SE)	Incremental Cost (SE)	95% CI
Bivariate	A 6668 (131.7)	161 (146.31)	(−125.9, 447.9)
	B 6507 (63.7)		
Univariate[a]	A 6668 (102.8)	161 (141.11)	(−115.7, 437.7)
	B 6507 (96.6)		

[a] Assuming normally distributed costs and fitting a simple linear model.

result in mean incremental costs of about £161. A MANOVA approach would report mean incremental differences (and effects).

In the univariate analysis, the standard error of the mean incremental cost is 141.11. For the bivariate model, the standard error of the mean incremental cost is higher, at 146.3. Therefore, the univariate analysis may have underestimated the uncertainty in the mean incremental effects (in this example) by not accounting for the correlation between costs and effects. The standard errors of the mean costs are higher for Treatment A and lower for Treatment B when compared with the univariate analysis; the sample standard deviations are 3,195 and 2,013, giving sample standard errors of 131.7 and 68.2 for Treatments A and B, respectively. The reader may wish to estimate the incremental effects using the same model as an exercise.

To compute the INMB, we need an estimate of the incremental effect for Treatments A and B. To calculate the INMB for a given λ = £20,000, the mean INMB is determined from INMB = $\Delta_e^* \lambda + \Delta_c$, giving INMB = Δ_e^*£20,000 + 161 (we just need to estimate the mean incremental effect). Although the option in SAS type = UN does not impose a specific covariance structure, if a new treatment lacks efficacy, resulting in higher costs because patients stop the new treatment for more complicated procedures, a suitable form of correlation could be stated in the model.

4.5.3 Transforming Cost Data

If costs are not normally distributed, a transformation might be considered. One common transformation is logarithmic (natural logs), when data are positively skewed and the variance of costs increases with its mean. The back-transformation of the average costs in terms of geometric means provides an estimate of the population median (not the mean). Therefore, unless a transformation provides unbiased estimates of the sample mean, for the purposes of cost-effectiveness analysis, it may not be useful in practice. Other problems with log transformations include zero values, which are undefined on the log scale. It is possible to add a small (but arbitrary) negligible value to the zero, but this does not overcome the concern that back-transformed values are not estimates of the sample mean and may perform

TABLE 4.7

Possible Transformations

Some Examples of Box–Cox Transformations		
Transformation Value (Power)	Resulting Value	Practical Uses
0.5	$Y^{0.5} = \sqrt{Y}$	For Poisson (count data) for stabilising the variance (e.g. makes smaller numbers larger and stabilises larger values; could be used for modelling resource use)
0	$\log(Y)$	When there are some very extreme costs (variance stabilising)
1	$Y^1 = Y$	Original response
−1	$Y^{-1} = 1/Y$	When there are some very extreme costs (variance stabilising)

poorly (Duan, 1983). An unbiased estimate of the population mean after log transformation is

$$\mu = \exp[\bar{x} + s^2 / 2]\,\bar{x}$$

where $\bar{x}$ and s^2 are the mean and variance of the log (cost) values on the log scale. A back-transformation of the standard error (needed for a probabilistic sensitivity analysis) can be far more complicated to derive. Moreover, there is no guarantee that even after log transformation the distribution of costs will yield a normal distribution, in which case the expression shown may still give biased estimates of the mean. Other transformations (e.g. cube, square root and reciprocal) forming part of the Box–Cox class, which attempt to achieve symmetry in the error, can also be used (see Sakia [1992] for a discussion of power transformations). The primary purpose of these transformations is to ensure that the assumptions of a linear model (usually assumptions around constant variance) hold true (Table 4.7).

4.6 Using Generalised Linear Models to Analyse Cost Data

In Section 4.5.2, we provided an example of a linear regression model which was used to analyse costs. In that example, one assumption was that the response variable (cost) was normally distributed and the variance of the costs was constant. The fact that costs are >0 may lead to a violation of normality assumptions, and, moreover, the variance of costs is likely to increase with higher expected costs (variance is a function of the mean and is not constant). Although the linear model might still be used, for analysing costs it may not be optimal. The OLS-type regression can be extended to a general class of linear regression models, GLM, which allow for situations where departures from normality can be handled efficiently. A fuller description of GLM can be found in Appendix 5C. We present a very brief overview of the concepts with the emphasis on practical interpretation of the results.

GLM can be used to model any response (binary, continuous, rate). The three main components of a GLM are

(i)   The random component (the costs of each patient are independently distributed, from one of the exponential family of distributions, such as normal, Poisson or gamma).

(ii) The systematic component (the explanatory variables, such as treatment, country, age, etc.).

(iii) The link function (the relationship between the systematic and random components). For responses which are normally distributed, the link function is often set to unity (constant variance); whereas if the distribution of costs is gamma, the relationship between the mean of costs and variance is expressed as a log function.

In practice, the explicit link function does not need to be stated, because it can be derived using software packages such as STATA and SAS. However, we do need to have some idea of the basic mathematical transformation. For example, for a gamma distribution, the mathematical transformation would be in terms of logarithms. Table 4.8 shows typical distributions with associated link functions used in SAS. Note, costs are unlikely to be binomial, although resource use (GP visit: Yes/No) might require a logit link function. One warning regarding GLM is that it may be prone to model misspecification (Glick, 2014).

Example 4.4: GLM Where Costs Are Considered Gamma Distributed

In this example we assume that costs are gamma distributed, such that the variance of costs is assumed to be proportional to the square of the mean (note: for a Poisson distribution, the variance is proportional to the mean). First, we plot the data and perform some preliminary tests of goodness of fit. Figure 4.4 shows the distribution of costs with various goodness of fit tests for normal (flatter curve) and gamma distributions

TABLE 4.8

Possible Link Functions for GLM

Distribution of Costs	Link Function
Normal	Identity (value = 1)
Poisson	Log
Gamma	Log
Binomial	Logit

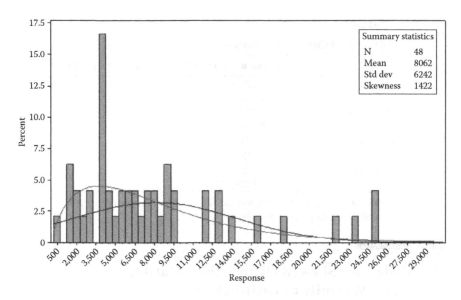

FIGURE 4.4
Normal and gamma fit to costs for data in Example 4 4.

(skewed curve). The assumption for normality is rejected (p < .01), whereas for gamma it holds (p > .25). The measure of skewness of 1.42 shows lack of symmetry.

Using a fixed effects GLM with age as a covariate for the data in Table 4.2, this SAS code fits a model assuming cost data are gamma distributed:

```
PROC GENMOD data =<dataset>;
Class treatment;
MODEL cost =treatment age/ DIST=GAMMA LINK=LOG;
LSMeans treatment / pdiff cilink;
Run;
Output:
```

The relationship between the mean of costs and its variance is expressed through the LINK = LOG statement. However, the output for the estimates of mean costs is not appropriate for an economic evaluation, because it is on the log scale. Therefore, we need to back-transform the output to provide the relevant difference in mean costs after adjustment for age (using the CILINK option). To compare the results of different models, we present the output in Table 4.9.

One important observation from Table 4.9 is that the incremental cost changes depending on the assumptions of the distribution (it decreases by 28% with the gamma assumption). In addition, the standard errors for the OLS are higher than for a GLM. It is these standard errors that would be used in any PSA. The standard errors on the raw scale are obtained through the CILINK option in the above SAS code for each of the mean costs.

TABLE 4.9

Incremental Cost for Normally and Gamma Distributed Costs

Model	Treatment Mean (SE)	Incremental Cost (£)
OLS	A 7792 (1290)	542
	B 8333 (1290)	
GLM: Normal	A 7792 (1250)	542
	B 8333 (1250)	
GLM: Gamma[a]	A 7798 (1243)	389
	B 8187 (1274)	
GLM: Log normal	A 7704 (1137)	630
	B 8335 (1230)	

[a] The LSMeans are back-transformed. The standard errors for the INB need to be computed manually.

4.7 Models for Skewed Distributions Outside the GLM Family of Distributions

The (fictitious) data in Example 4.4 are one example of how patient-level cost data might be analysed. However, when the data present with complications, such as many zero costs (right skew), there are several alternative choices of models which can be used. Some of these models are known as *two-part models* (Grootendorst, 1995).

Two-part (or hurdle) models might be used when excess zeros are present, that is, when many patients do not use healthcare resource. An example of this might be instantaneous deaths at randomisation or just prior to the start of study treatment. In this case, a model is first used to estimate the chance of resource use (using a logistic or probit-type model), and then a second model is used to compute the expected (mean) intensity of the resource use. For example, if there were a considerable proportion of patients with zero costs, fitting a (linear) straight line model to the data in the graph in Figure 4.5 would be misleading because of the many zero values on the line $y = 0$ (costs = 0).

In a two-part model, a logistic model might first be fitted to estimate the probability of resource use >0 for each patient, and then a separate model (such as an OLS, GLM or other more complicated model) might be fitted, conditional on the first part. For example, the probability of hospitalisation

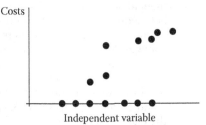

FIGURE 4.5
Relationship between costs and covariates with zero costs.

is modelled using a logistic model, and the second part could then model the intensity (duration) of the hospitalisation, as in Grootendorst (1995) and Pohlmeir & Ulrich (1995), where a zero-inflated negative binomial model was used in the second part to estimate mean costs. In PROC GENMOD in SAS, this can be accomplished using the 'DIST = Zinb' option.

The main advantage of using two-part models is that they avoid biased estimates of incremental costs associated with situations, as in Figure 4.3. In some disease areas, from a clinical trial perspective, zero costs are unlikely to occur. In cancer, for example, patients enrol onto the trial with the general view that they will receive treatment in one form or another. If many patients were not receiving treatment (zero drug costs) due to early death (withdrawal from the trial), this might indicate the wrong inclusion criteria or patient population. One consideration for two-part models is that the models may involve two separate stochastic processes. The decision to visit the GP and the intensity of the GP visit, for example, are based on two separate decision processes: one with the patient, the latter with the treating physician. Ignoring these differences can lead to potential misinterpretation (Pohlmeir & Ulrich 1995). Although these models are used in modelling economic data, their use in a clinical trial context might be limited. A Bayesian approach has been applied to clinical trial data (Baio, 2014).

4.8 Summary of Modelling Approaches

Mihaylova et al. (2011) provide a summary of 17 different modelling or analysis approaches for cost data. The methods described in this chapter can be considered to be those that are relatively tried and tested. The authors place these in a so-called Green and Amber zone – suggesting that if the assumptions are satisfied, these methods can be used relatively safely. More complex approaches, such as Markov Chain Monte-Carlo (MCMC) methods and those which incorporate censoring of costs directly or model over different assessment points throughout the trial, are also possible. For example, higher costs earlier in the intervention and lower costs later might be observed in surgery trials, because the hazard for disease progression and death might change immediately after surgery compared with a later stable period.

One suggested conclusion on the use of the various modelling approaches presented is that the simplest approaches, such as OLS models, are generally suitable when sample sizes are sufficiently modest to large and the skewness is not extreme. If the degree of skewness (Manning & Mullahy, 2001) is substantial, GLM with a gamma distribution and log link function will provide better precision than OLS without a log transformation. However, compared with an OLS model in which costs are log transformed, GLM assuming a gamma link function will provide less precise estimates, although the estimates of the mean costs for each treatment group (and the cost difference) will be similar.

Transformation of data may be a possible approach, depending on the ability to back-transform on the raw scale to derive mean cost differences between treatments. Indeed, it is important that whatever model is used, the mean cost difference and its corresponding standard errors on the raw scale are available for later use in sensitivity analysis. When costs are zero, adding small values is not recommended. When excess costs are zero, two-part models are particularly preferred. More complex methods are evolving (e.g. Baio, 2014), and time will tell as to the relative superiority of these newer methods over established approaches. If the gain is minimal, simple approaches are always recommended.

4.9 Handling Censored and Missing Costs

We distinguish between two types of missing data:

(i) Missing in the sense that the patient missed the visit, the CRF page was lost or the health resource was used, but the quantity was unknown (e.g. the data of hospital admission were recorded in the CRF, but the duration was unknown). Another example might be where resource use CRF pages are lost, but the proportion lost is similar for each treatment group. An important issue for this type of missing data is that missing is not the same as zero cost, and a resource use set to zero is not the same as saying resource use is missing.

(ii) An incomplete record of health use because either the patient was lost to follow-up or the study was 'complete' before the primary outcome was observed. This type of missing data is known as censored data (censored at the last date of contact). In this case, the total costs for that patient represent a minimum (because if followed up, the costs might be larger). Consequently, calculating a simple mean cost would be biased. It has been suggested (Willan & Briggs, 2006) that patients who are censored will have smaller costs, and therefore the sample mean cost will be an overestimate.

One reason for missing resource use (cost) data is when patients withdraw from the trial early, either due to lack of effect, adverse events or early death (due to a competing event) or when the study has ended before the outcome of interest can be observed. Missing data are problematic in any clinical trial, whether the outcome is resource use or any other clinical end point, and is more likely to occur in clinical trials in which the primary end points may be reached several years after randomisation (e.g. survival end points). There is often little or no plan to collect any additional data after study withdrawal or adverse events.

Missing data of type (i) can be handled through well-established analysis methods such as multiple imputation and other complicated approaches (e.g. shared parameter models) involved in testing assumptions about the mechanism of missingness, such as missing completely at random (MCAR), missing at random (MAR) and missing not at random (MNAR). A detailed discussion of these would require a separate volume, and they will not be elaborated on further here. References in the bibliography can be consulted by interested readers (e.g. Carpenter & Kenward, 2013; Farclough, 2010). Most literature on missing data in economic evaluation tends to focus on censored costs. The focus here is less on estimating missing health resource (e.g. number of GP visits for a particular patient) than on the monetary value itself of the missing resource. Therefore, attention will be paid to missing data of type (ii) for the purposes of deriving the incremental costs.

In many clinical trials, the intention to treat (ITT) population is often the primary population for analysis. This means that regardless of any violations, compliance or early withdrawal from the trial, if the patient was randomised, his or her data should be analysed. If data are missing at some point after randomisation or midway through treatment, strategies are needed to deal with how to handle such missing data. What makes this issue particularly relevant to the analysis of costs is that patients who have missing data might well be the same patients who also have higher costs (because the missing values might be associated with problems with side effects and, ultimately, the treatment). Patients who drop out from the trial due to toxicity from the experimental medicine may go on to receive additional medication or treatment for their adverse events. If the patient has been lost to follow-up, the costs are likely to be underestimated. Although it is difficult for patients to recall their pain response over the last 7 days (as is often done in pain studies), they are unlikely to forget that they went to visit the GP or saw a nurse. Therefore, if 'Yes' was ticked on the CRF to a question regarding whether a GP visit had been carried out, but the date of the visit was missing, this would not necessarily be considered as a missing data component.

4.10 Strategies for Avoiding Missing Resource Data

There are several approaches to preventing missing or censored cost data:

- Design clear and easy-to-understand CRFs when collecting resource use, with clear instructions.
- Think carefully whether assessment time points are adequate. Assessments which are too far apart, for example 1 month and 6 months, may result in costs being missed between these months, and also the recall period may be too long. For example, a question

at 6 months 'Did you have any other medication for your pain since your last assessment?' requires the patient to remember everything that happened between the first and the sixth month. It is better to have a tick box with options.

- Continue to collect all or some data after the patient has discontinued from the study or has discontinued from the trial (if possible).
- Select outcomes which are easily ascertainable.
- Use those clinical sites which have good experience with following up (not just the ability to recruit).
- Train trial staff on how to communicate with the site staff so that follow-up data can be obtained. It is not helpful when a rude trial coordinator or clinical research associate (CRA) demands that site staff obtain the missing data. There will not be much motivation to obtain follow-up data if communication is a problem.
- Pay site investigators for quality of follow-up and completeness of CRF forms, or use other forms of incentives.
- Inform patients why collecting resource use is important: because it will help determine whether the new experimental treatment offers value for future patients (this may not go down too well with all patients!).

4.11 Strategies for Analysing Cost Data When Data Are Missing or Censored

Several strategies are suggested for analysing data in the presence of missing or censored cost data. There are several general approaches for handling missing or censored data at the analysis stage:

(a) Complete case analysis

(b) Imputation methods

(c) Model-based methods

When there is complete cost data, any method can be used to analyse the costs. If patients who do not have complete cost data are excluded from the analysis, this will not only violate the ITT principle, but it will also lead to biased estimates of mean costs (Huang, 2009; Wijeysundera et al., 2012; Zaho, 2012). Although data from patients who are censored are omitted, leading to loss of information, the associated loss of power for statistical inference is less of a practical concern, because differences in mean costs are not powered for statistical significance. What is more important is the resulting bias in the estimate of the mean and standard errors of the incremental cost.

4.12 Imputation Methods

Single imputation methods such as last observation carried forward (LOCF) should be avoided. Just as, with clinical end points, the LOCF is not realistic when the disease becomes more severe over time, similarly, with cost data, the costs may increase over time as the disease progresses. Other imputations, such as using the mean costs, worst case or baseline carried forward, which are sometimes used for clinical end points (and suggested in some clinical regulatory guidelines; EMEA [2010]), also result in biased estimates of mean costs and are not appropriate for costs. For example, the worst case for a clinical end point involving pain, on a scale of 0 to 10, might be 10, but the 'worst' or 'maximum' cost could be anything; similarly, baseline carried forward might be a way to determine conservative treatment effects, but baseline costs can be lower, because on follow-up, as the disease progresses, costs increase.

Multiple imputation methods, on the other hand, might be suitable for handling some type (i) missing methods to improve estimates of the mean cost. While the single imputation approaches are effectively (poor) guesses for the data that are missing, multiple imputation is a slightly more complicated method, which results in a better guess than other methods (e.g. worst-case scenario). These methods are discussed extensively elsewhere (Rubin, 1987; Carpenter & Kenward, 2013). To generate the mean incremental cost when using this approach, an imputation model is assumed, and the missing costs are predicted several times (several complete data sets are generated). Subsequently, the mean cost is determined for each treatment group for each of the data sets. Finally, algorithms (Carpenter & Kenward, 2013) are used to derive the final point estimates of mean costs for each treatment group, and the mean incremental cost is derived.

4.13 Censored Cost Data

In contrast to the simple imputation methods described in Section 4.12, two particular approaches will now be discussed, neither of which shows definitive evidence of being better, but both of which are acceptable methods for handling cost data that are censored. Where censoring is of concern (as in mortality end points), patients who live longer are more likely to be censored, and the bias is in the direction of those patients who die earlier (underestimate of mean costs). Wijeysundera and colleagues (2012) have compared some of these methods with newer approaches. We introduce practical examples of these.

4.14 Method of Lin (Lin et al. 1997)

This method provides an estimate of the mean cost by taking into account patients who are followed up to a particular time point and then are lost to follow-up (censored). The estimate of mean costs uses the survivor function (Kaplan–Meier) to generate weights, which are multiplied by the mean costs for specified intervals. For trials that do not have a mortality end point, the event of interest might be time to drop out. In a 12-week study where mortality is not the primary end point of the study, almost all patients are likely to remain in the study, and most patients would be censored.

The calculations for mean costs can be broken down into the following steps:

1. Divide the follow-up period into smaller (not necessarily equal) time intervals. For example, a follow-up of 1 year could be split into 12 equal intervals (1 month apart).

2. Calculate the mean costs within each interval (i.e. for each of the 1 month intervals) only for those patients alive at the start of the interval. For example, if 100 patients are alive at the start of month 1, then the mean 1 month cost will be determined for the 100 patients.

3. Compute the survival rates at each interval (every month in this example) using survival methods (typically Kaplan–Meier plot). For example, the overall survival rates at 1 month might be 90% (after 1 month, 10% have died).

4. Calculate the expected monthly cost by multiplying the monthly survival rates by the monthly costs.

5. Add up the mean costs for each of the 12 months (i.e. month 1 + month 2 + ...month 12) to get the total mean costs (total of the means).

6. Repeat for each treatment group.

One limitation of this method is that it does not adjust for covariates when estimating the mean cost. However, Lin (2000) provides an extension of this, which can adjust for covariates.

Example 4.5: Lin's (1997) Method

Table 4.10 shows costs and mean costs calculated for each treatment over a 12 month follow-up period using the Lin (1997) approach. First, we split the 12 months into 2-monthly intervals and derive the costs (treatment costs, nurse visits, etc.). One might think of each interval as a cycle of 8 weeks (roughly 2 months) of treatment for a particular cancer, corresponding to months (0–2), (2–4), (4–6), (6–8), (8–10) and (10–12).

In Table 4.10, the mean costs are computed for each interval, taking into account whether patients have complete cost data in the interval.

TABLE 4.10

Adjusting for Censored Costs

Patient	Treatment	Interval					
		1 (0–2)	2 (2–4)	3 (4–6)	4 (6–8)	5 (8–10)	6 (10–12)
1[a]	A	5000	3000	4000	7000	6000	5000
2	A	7000	6000	2000	300	0	0
3 etc.	A	etc.	etc.	etc.	etc.	etc.	etc
	Mean A	6300	4900	4200	3650	4520	5450
	Treatment	**1 (0–2)**	**2 (2–4)**	**3 (4–6)**	**4 (6–8)**	**5 (8–10)**	**6 (10–12)**
51[b]	B	7000	2000	0	0	0	0
52	B	5000	3000	4000	8000	4000	0
53 etc.	B	etc.	etc.	etc.	etc.	etc.	etc
	Mean B	5300	3900	1200	4650	6170	3195
$S(t)_A$	A	90%	85%	80%	40%	20%	10%
$S(t)_B$	B	90%	75%	65%	20%	5%	2%

[a] Patient censored.
[b] Patients 2 and 51 died during intervals 4 and 2, respectively.

Costs calculations are based on patients alive at the beginning of the interval. The survival rates are obtained by plotting the survival times (assuming the end point is mortality) and using the observed product limit estimates. For each group, then, the cumulative (i.e. total) mean cost for Treatment A is computed as $(£6,300 \times 0.9) + (£4,900 \times 0.85) + (£4,200 \times 0.8) + (£3,650 \times 0.4) + (£4,520 \times 0.2) + (£5,450 \times 0.1) = £12,744$. The same calculation is then performed for Treatment B, and then the incremental cost can be calculated. Note that this method is based on patients alive at the beginning of the interval. An alternative method is also available (Lin, 1997, method 2), which computes the mean costs of those patients who die within an interval and then multiplies these costs by the probability of death within the interval (from the Kaplan–Meier survival estimates). Adjustment for covariates when estimating mean costs in the interval is also possible (Lin, 2000). Whatever method is used, care should be taken to check the assumption that censoring within the interval is MCAR.

In the methods described in Section 4.14, Lin (1997) assumes that the censoring occurs at the beginning or end of an interval. To take into account situations in which censoring is continuous, Bang & Tsiatis (2002) use an inverse probability weight (IPW) to estimate mean costs. The weighting is determined by the reciprocal of the survival probability (1/S(t)), such that patients who die early are given lower weight. This is less efficient, because it only uses complete observations.

Table 4.11 shows the details of the types of costs for drugs and where they can be obtained in some countries.

TABLE 4.11

Some References for Obtaining Data on Sources for Unit Prices

Country	Sources for Costs/ Healthcare Resource	Website (Last Access Date)[a]
United Kingdom	BNF	www.bnf.org (8 April 2013)
	NHS Costs	https://www.gov.uk/government/collections/nhs-reference-costs
United States	Centers for Medicare & Medicaid services	www.cms.gov (8 April 2013)
Germany	Rote-list	http://online.rote-liste.de/Online/login
Spain	Vademecum	http://www.vademecum.es/ (8 April 2013)
Italy	L'informatore Farmaceutico	www.informatorefarmaceutico.it/ (8 April 2013)
Canada	Canadian Institute for Health Information	http://www.cihi.ca (8 April 2013)

[a] In some countries, access to the website can only be obtained when in the country and may require registration.

4.15 Summary and Conclusion

In this chapter, we have discussed several important and common aspects of modelling cost data. To cover the entire spectrum of modelling approaches would make this work unnecessarily long. These include Bayesian approaches (Lambert et al., 2008; Baio, 2014), random effects using two-part models (Liu et al., 2010), using the IPW approaches mentioned in Section 4.14 in combination with covariates (Basu & Manning, 2010) and non-parametric approaches (Zaho et al., 2012), among others. One feature common to the literature when analysing costs data is the issue of censored data and informative censoring, which render estimates of mean costs from traditional survival analysis methods invalid. The more complicated methods are not necessarily the 'best' approaches, and it is important to note that a key aspect of modelling in economic evaluation is not about hypothesis testing, but, rather, about quantifying the uncertainty in the parameters, in particular estimates of standard errors of the mean incremental costs.

Exercises for Chapter 4

1. What would be the consequences for the ICER if the costs were discounted but not the effects?

2. When is it more appropriate to model the health resources rather than modelling the total costs? Would it make a difference to the ICER?

3. Using the data in Table 4.2, estimate the mean costs, assuming the data are log normally distributed. Interpret your results and compare with the assumption that costs follow an alternative distribution.

4. Is the assumption of a normal distribution for costs appropriate?

5. How would you go about checking the distribution of costs and choosing a suitable approach to computing the mean costs?

6. Why would a reimbursement body such as NICE or the Scottish Medicines Consortium (SMC) be concerned about a large amount of missing resource data on the comparator treatment but not on the control treatment?

7. Discuss strategies for handling missing patient-level costs. Is imputation always appropriate?

Appendix 4A: SAS/STATA Code

4A.1 SAS Code for Modelling Costs

```
Proc glm data = <dataset>;
Class treatment;
Model Cost = Treatment;
Run;

**adjusting for centre or country effects**;

PROC MIXED data = <dataset>;
Class treatment country;
Model Cost = treatment country;
Random patient;
Lsmeans treatment;
Run;
```

4A.2 SAS Code for Joint Modelling of Costs and Effects

Generate data for bivariate regression example.

```
data simcost1;
do pat=1 to 1000;
c1=rannorm(576);
resp=c1*4518+5402;
treat='A';
outcome='Costs';
output;
```

```
end;
drop c1;
run;

data simcost2;
do pat=1001 to 2000;
c2=rannorm(5723);
resp=c2*2005+6544;
treat='B';
outcome='Costs';
output;
end;
drop c2;
run;

data costs;
set simcost1 simcost2;
if resp <0 then delete;
run;

data simeff1;
do pat=1 to 1000;
e1=rannorm(5735);
resp=e1*0.89+1;
treat='A';
outcome='Effects';
output;
end;
drop e1;
run;

data simeff2;
do pat=1001 to 2000;
e2=rannorm(5735);
resp=e2*0.442+0.967;
treat='B';
outcome='Effects';
output;
end;
drop e2;
run;

data effects;
set simeff1 simeff2;
if resp <0 then delete;
run;
```

4A.2.1 Bivariate Model

```
PROC MIXED data =<dataset>;
Class Treatment outcome;
```

```
Model Response = outcome treatment outcome*treatment;
Repeated outcome/ type=un grp=outcome*treatment sub=pat;
Lsmeans outcome*treat/cl;

Run;
```

4A.3 SAS Code for Modelling Costs: Using a Generalised Linear Model – Assuming Costs are Gamma Distributed

```
PROC GENMOD data =<dataset>;
Class treatment;
MODEL cost =treatment age/ DIST=GAMMA LINK=LOG;
LSMeans treatment / pdiff cilink;
Run;
Output:
```

5

Quality of Life in Economic Evaluation

In this chapter, the rationale for collecting quality of life (QoL) data will be discussed, including the controversies surrounding how they should be measured and whether disease-specific or generic health-related quality of life (HRQoL) measures are more suitable for economic evaluation. This is then followed by showing how a commonly cited generic measure in economic evaluations, the EuroQol EQ-5D-3L (EQ-5D), is used in practice, and estimating patient-level EQ-5D data using mapping functions when such data have not been collected in a clinical trial. The quality-adjusted life year (QALY) is then described, with some practical examples of how it is derived under various scenarios. Quality-Adjusted Time Without Symptoms or Toxicity (Q-TWiST) is also discussed, followed by a discussion of statistical and trial design issues relating to HRQoL in the context of an economic evaluation.

5.1 Quality of Life in Clinical Trials versus Quality of Life for Economic Evaluation

QoL is often a vague concept with a variety of meanings depending on one's point of view. Some interpret QoL as a general measure of well-being that captures (rather subjectively) differences in hopes and expectations between an individual's past and present experiences (Calman, 1984). Although QoL is hard to define exactly, we might have some general 'feeling' of what QoL is about. The QoL of an individual is affected by factors such as life–work balance, home life, family relationships and so on. QoL data are often collected in a clinical trial, when a new treatment might show efficacy to be improved or equivalent to the standard, but also offers additional benefit in the form of HRQoL (e.g. better safety profile leading to improved symptom control).

In clinical trials, we are really interested in measuring HRQoL, which attempts to capture the impact of treatment and disease on a patient's health. An HRQoL instrument is designed to measure a patient's physical function (through the severity of symptoms and toxicity), psychosocial function (well-being, distress, self-esteem), functional ability (self-care, shopping, work) or social functioning (relationships). In an economic evaluation, the use of HRQoL is particularly important for cost–utility analyses (CUA). If a cost-effectiveness analysis involves reporting the incremental cost-effectiveness ratio (ICER) (or incremental net monetary benefit [INMB]) in terms of costs per natural unit (e.g. cost per unit reduction in blood pressure), then HRQoL becomes less important for constructing the ICER; HRQoL can, however, still be used to demonstrate the value of a new treatment, because it is often an important secondary end point.

One reason why HRQoL is collected is because clinicians need to maximise information about how well (or not) drugs are working so that they can make informed decisions when treating patients. It is essential to know (from both patient and clinician perspectives) not only what side effects are associated with treatments, but also how treatment-related side effects impact patients' HRQoL. It is now accepted that HRQoL should be measured in clinical trials, but the debate continues as to the most reliable and practical way of obtaining these data (Slevin, 1988). Some have suggested that a 'treatment can be recommended ... even without an improvement in survival, if it improves quality of life', particularly for diseases such as cancer (ASCO guidelines, 1996). There are also several surveys that suggest that HRQoL can influence the choice of treatment in several disease areas. In 8% of randomised controlled trials (RCTs) in breast cancer, HRQoL influenced a treatment decision (Goodwin, 2003); in 25% and 60% of prostate cancer and surgical trials, respectively, the treatment decisions were influenced by HRQoL (Efficace et al., 2003; Blazeby et al., 2006). One other important aspect is that HRQoL should be collected directly from patients (patient-reported outcomes) and not through proxies (such as the carer, clinician or nurse).

5.2 Disease-Specific and Generic Measures of HRQoL

HRQoL has until relatively recently been restricted to condition-specific measures (CSM) in clinical trials. A CSM is an instrument that captures specific QoL issues in patients with a given disease. For example, if a patient has cancer, questions related to their cancer are asked (e.g. using an instrument such as the EORTC-QLQ-C30); or if the patient has arthritis, then the Health Assessment Questionnaire (HAQ) is completed. The EORTC-QLQ-C30 (used an example in this and some later sections) is a 30-item questionnaire consisting of 15 domains, scored on a 0–100 scale. The scoring consists of five function scales: physical function (PF), role function (RF), emotional function (EF), cognitive function (CF) and social functioning (SF). There are also nine symptom scales: fatigue (FA), nausea and vomiting (NV), pain (PA), dyspnoea (DY), insomnia (IN), appetite loss (AL), constipation (CO), diarrhoea (DI) and financial problems (FI); there is also a global health status score (QL). For the global health and function domains, high scores indicate better QoL. For the symptom domains, low scores indicate better symptoms. The QLQ-C30 is used to measure HRQoL in several cancers.

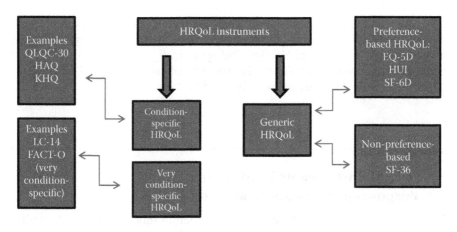

FIGURE 5.1
Example of the relationship between generic and condition-specific HRQoL measures.

In addition to CSMs, there is another group of HRQoL measures that are generic across disease areas (e.g. the short form 36 [SF-36], SF-6D, the health utilities index [HUI] and the EQ-5D). Within this group of measures, however, only some are used for effectiveness analyses. The HRQoL questionnaires used for economic evaluations are called *preference-based measures*. Figure 5.1 shows the relation between generic and preference-based measures.

Although responses from the QLQ-C30 reflect how a given patient might feel with regard to his or her symptoms, disease or even treatment received, the responses do not necessarily reflect how payers (people who ultimately pay for treatments through taxes) perceive the value of such a patient's health condition, even if the patient does have a disease as serious as cancer.

It is possible that society (not necessarily doctors) might regard another patient's health (state) as being far worse, and therefore believe that resources available to treat a patient's illness should be used elsewhere (e.g. preference for a breast cancer sufferer over a lung cancer patient). Condition-specific instruments do not incorporate a relative *valuation* of how much a specific symptom affects the overall perception of health. For example, severe nausea might be considered worse than severe pain for a given patient. A CSM does not usually easily incorporate this *relative preference* or value of a given health state (although there is at least one example of a CSM adapted for utility elicitation: Rowen, 2011). Utility is a measure of such a preference for a given health state measured on a scale from 0 (worst imaginable health) to 1 (perfect health). States considered worse than death have values (unsurprisingly) below 0.

Instruments such as the QLQ-C30 and and the King's Health Questionnaire (KHQ) are not *preference*-based measures of HRQoL, and therefore are not usually used directly in an economic evaluation.

In order to make a decision about value or preference to treat patients with specific health technologies, a different HRQoL instrument is used. This is a very generic measure of HRQoL and specifically developed for use in an economic evaluation. The EQ-5D is one example of this type of preference-based instrument. Others include the HUI and the SF-6D.

5.3 HRQoL Instruments Used for the Purposes of Economic Evaluation

Two principal methods can be used to generate (elicit) health-state utility scores for use in an economic evaluation: direct and indirect valuation. Direct valuation, such as time trade-off (TTO), involves asking individuals (or patients) to 'trade off' or 'give up' certain amounts of a particular health benefit, such as longer survival time, for more (or less) of an alternative health benefit. For example, one might be asked to express a preference between '12 months of survival with HRQoL valued at 0.6' and '14 months of survival in a HRQoL of 0.45' (0 = death, 1 = perfect health). The preferences for a given health state are elicited directly. The data collected from these questions are subsequently analysed, and weights are derived to determine a utility index. An example of this type of elicitation in breast cancer patients is reported by Sime and Coates (2001).

Indirect elicitation, on the other hand, essentially uses HRQoL questionnaires in which the raw responses from the instruments are converted into utilities using a standard set of predefined weights (based on direct elicitation methods). These are also called multi-attribute instruments, from which health states are determined and then the aforementioned preference weights are applied. The meaning of *health state* will now be elaborated in the context of the EQ-5D.

5.3.1 EQ-5D: Generating Health States and Applying Preference Weights

The EQ-5D is a generic HRQoL instrument recommended for use in economic evaluations by the National Institute for Health and Care Excellence (NICE) (Brazier, 2011). Details of the EQ-5D are well documented (Dolan, 1997; Mann, 2009). The EQ-5D consists of a descriptive health-state classification system with five domains (mobility, self-care, usual activities, pain/discomfort and anxiety/depression) and three severity levels in each (no problems, some problems and extreme problems). Combining one level from each domain defines 243 different health states, ranging from full (value of 1) to worst (value of 0) health.

For example, a score of 2 for mobility, 1 for self-care, 3 for usual activities, 3 for pain/discomfort and 3 for anxiety/depression defines a health state of 21333. Values of 11111 (a score of 1 for each question) define the health state '11111', the 'best' or 'perfect' health state, while the poorest (worst) health state is 33333. Each of these health states is converted to a single number: a value from −0.594 to 1 (1 is full health and −0.594 represents the worst state possibly imaginable – even worse than death). Values such as 11111 or 21333 are not easily analysed, but when they are converted to a single value, this can be analysed. More recently, the EQ-5D-5L has scores ranging from 1 to 5 (for each of the five domains). Hence, while a perfect health state is '11111', the worst health state is now '55555'.

Example 5.1: Converting EQ-5D Scores to Utilities

A patient presents with raw scores from the EQ-5D of 2 for mobility, 1 for usual activities, 3 for pain/discomfort, 2 for self-care and 2 for anxiety/depression. This combination yields a health state of 21322. From the published weights given by Dolan (1997) using the TTO method, this patient's utility is 0.293. It is this value that will be subsequently applied to clinical measures such as overall survival (OS) or progression-free survival (PFS) time to derive a QALY. An algorithm to convert raw EQ-5D is available in Brazier et al. (2009), and SAS code is also provided in Appendix 5A.1.

There are likely to be differences between patients and the general population in how a given health state is valued. Regan (1991) observed notable underestimates of the HRQoL in the general population compared with patients. The main argument for generic measures is the ability to compare across healthcare programmes, even when treatments and medical conditions differ. Some research suggests that CSMs are inappropriate or insensitive for many medical conditions (Harper et al., 1997; Kobelt, 1999; Barton, 2004), and using an existing and established generic measure may also improve its acceptability to the clinical community. Moreover, there is a concern that CSMs can also capture the HRQoL effects of co-morbidities, which may bias response.

Some (but not all) generic measures are easier and shorter in duration for patients (the EQ-5D has only five questions). This results in less missing data and consequently improved precision around any estimates of measures. There is still some question as to whether a generic measure is, indeed, less sensitive, and, if it is, how the 'loss' in sensitivity should be quantified (Bozzani et al., 2012 and Khan, in press also discuss this to some extent).

The summary of the debate between CSMs and generic measures is that CSMs are more sensitive and cover more important aspects of the condition that are missed using generic instruments. Preference-based generic measures are generally suitable for cost-effectiveness evaluation for deriving cost per QALYs. Consequently, QALYs can be used to compare the value of several treatments in different disease areas.

5.4 When HRQoL Data Have Not Been Collected in a Clinical Trial

5.4.1 Mapping Functions

One method to determine utility when a generic HRQoL has not been used in a trial involves a separate utility study (direct elicitation), as described in Section 5.3. The alternative approach is to determine utility values using a mapping function or 'cross-walking'. Mapping is a method by which the interrelationship between a generic HRQoL measure such as the EQ-5D and a condition-specific HRQoL measure (e.g. EORTC-QLQ-C30) is modelled so that utilities can be predicted in studies in which the generic measure was not used. A key objective is to use the estimates of utility from the model for an economic evaluation, often for computing QALYs.

Mapping can be useful when a measure of utility is not available or when it is not planned to collect this information in a clinical trial. A statistical model, termed a *mapping algorithm*, is used to predict the EQ-5D from a disease-specific measure such as the QLQ-C30. If patient-level utilities cannot be obtained, then reliance is placed on published aggregate utility values, which have additional uncertainties (such as differences in trial design, patient populations etc.). Mapping may, therefore, be the only way to estimate individual patient-level utilities for a trial and can avoid the potential biases and uncertainties associated with using published aggregate utilities.

In some situations, the observed mean difference between treatments using the EQ-5D may not be expected to be very large compared with a CSM. In this case, the EQ-5D might not be used in the trial for fear that it might provide serious underestimates of HRQoL differences. If a useful mapping function was available, EQ-5D could be predicted in the hope that more realistic estimates of mean utility, comparable with those from a CSM, could be obtained. On the other hand, if large discrepancies are expected between the EQ-5D and a CSM, using a mapping algorithm may not be sensible (because there should ideally be some overlap for mapping to work well). In any case, it is not recommended to avoid collecting patient-level utilities, because mapping is a 'second-best' option for deriving patient-level utilities.

Example 5.2: Using a Published Mapping Algorithm

A mapping algorithm given by McKenzie et al. (2009) that predicts EQ-5D from the QLQ-C30 is applied to the data in Table 5.1. The objective is that, given patient-level QLQ-C30 data from the 15 domains (i.e. a QLQ-C30 score for each of the 15 domains for each of the patients), we can predict the patient-level EQ-5D utilities from a published algorithm using a simple linear predictive relationship:

$$\text{EQ-5D} = 0.2376 + \text{SF} \times (0.0002) + \text{RF} \times (0.0022) + \text{PF} \times (0.0004) + \text{EF} \times (0.0028)$$
$$+ \text{CF} \times (0.0009) + \text{FA} \times (-0.0021) + \text{NV} \times (-0.0005) + \text{PA} \times (-0.0024)$$
$$+ \text{DY} \times (0.0004) + \text{IN} \times (0.00004) + \text{AL} \times (0.0003) + \text{CO} \times (0.0001)$$
$$+ \text{DI} \times (-0.0003) + \text{FI} \times (-0.0006) + \text{QL} \times (0.0016)$$

Hence, for Patient 1 in Table 5.1, the predicted EQ-5D utility is estimated as

$$\text{EQ-5D} = 0.2376 + 50 \times (0.0002) + 63 \times (0.0022) + 81 \times (0.0004)$$
$$+ 5 \times (0.0028) + 33 \times (0.0009) + 55 \times (-0.0021) + 66 \times (-0.0005)$$
$$+ 33 \times (-0.0024) + 33 \times (0.0004) + 47 \times (0.00004) + 66 \times (0.0003)$$
$$+ 74 \times (0.0001) + 31 \times (-0.0003) + 29 \times (-0.0006) + 88 \times (0.0016)$$
$$= 0.447.$$

The process is repeated for all patients, and the means and standard errors of the predicted EQ-5D utilities are used for further analysis (inputs into the health economic model). In practice, several algorithms might be used for sensitivity analyses (i.e. how sensitive is the ICER to different predictions of the EQ-5D from different mapping models?).

It is not uncommon to find a mapping algorithm being used to predict patient-level EQ-5D for a clinical trial in a particular disease area (e.g. pain), although the mapping algorithm was developed using data from quite a different patient population. Crott and Briggs (2012) suggest that different algorithms or functional forms may exist for each cancer type in the case of QLQ-C30, and a single mapping algorithm across all cancer types may not be possible; a similar concern was raised also by Brazier et al. (2010). Therefore, it is not often clear whether published mapping algorithms intend users to extrapolate to *any* patient population, or how timing of measurements or presence of differential treatment effects can influence predicted values. For example, where algorithms have been developed using only baseline data, it is not immediately clear how useful they are for predicting post-baseline EQ-5D; users of algorithms

TABLE 5.1

Example of Using a Published Mapping Algorithm to Compute EQ-5D

Patient	SF	RF	PF	EF	CF	FA	NV	PA	DY	IN	AL	CO	DI	FI	QL
1	50	63	81	25	33	55	66	33	33	47	66	74	31	29	88
2	30	43	61	35	33	25	68	33	39	47	26	14	51	33	38
3	etc.														

AL: appetite loss; CF: cognitive function; CO: constipation; DI: diarrhoea; DY: dyspnoea; EF: emotional function; FA: fatigue; FI: financial problems; IN: insomnia; NV: nausea and vomiting; PA: pain; PF: physical function; QL: global health status score; RF: role function; SF: social functioning.

are often interested in predicting differential (post-baseline) treatment effects in terms of EQ-5D for showing cost-effectiveness.

Example 5.3: Developing and Testing a Mapping Function: A Non-Linear Beta Binomial Regression Model for Mapping QLQ-C30 to the EQ-5D

In this example, we used data from two separate non-small cell lung cancer clinical trials (TOPICAL and SOCCAR: Lee, 2012 & Maguire, 2013) to develop and validate a beta binomial (BB) mapping model. More details about the model can be found in Khan & Morris (2014). We also compared this novel BB model with commonly used linear algorithms as well as less common algorithms such as TOBIT, quantile, quadratic and censored least absolute deviation (CLAD) models. The data used were from two national (UK) non-small cell lung cancer clinical trials. The first trial (TOPICAL) was a randomised Phase III trial comparing OS between Erlotinib and Placebo in 670 non-small cell lung cancer patients. Both the EQ-5D and QLQ-C30 were collected monthly until death. The analysis was based on using data over the first 12 months (because nearly all patients had died by then). The SOCCAR trial was a Phase II randomised non-small cell lung cancer trial comparing sequential chemotherapy followed by radical radiotherapy (experimental arm) with concurrent chemo-radiotherapy followed by chemotherapy in 130 patients with inoperable Stage III non-small cell lung cancer. The EQ-5D and QLQ-C30 data were also collected monthly for a period of at least 18 months. Details of baseline characteristics have been reported elsewhere (Maguire, 2014).

Model Specification

For each of the models reported, data were combined across time points and treatment groups, so that results can be compared with previous approaches. One reason for pooling across all time points is that more health states can be modelled. For linear models, an underlying linear relationship between EQ-5D and QLQ-C30 was assumed, with constant variance of the error term. Patients were considered as random effects to allow for clustering within patients. In all models, adequacy of fit was considered using residuals, tests for homoscedasticity and Akaike's Information Criterion (AIC), where appropriate.

The models tested included

  (a) Linear ordinary least squares (OLS) model
  (b) Quantile regression
  (c) CLAD
  (d) BB regression

The BB distribution is often used in probabilistic sensitivity analyses (PSA) in health economic modelling for utility measures such as the EQ-5D. The reason for use in PSA is usually the convenience of assuming a scale from 0 to 1 for utility. In particular, the BB regression can model responses on the 0–1 scale that are unimodal or bimodal with varying levels of skewness. Kharroubi et al. (2010) suggest that utilities are likely to be skewed and truncated; therefore, the BB approach may be a suitable model to test for developing a mapping algorithm.

The BB regression has reported superiority over the OLS approach in terms of more accurate and efficient parameter estimates, particularly when the response variable follows a skewed distribution. An important feature of this approach is that mean predicted estimates of EQ-5D can be estimated while restricting the range to between 0 and 1. A response variable (EQ-5D) is assumed to follow a beta (α, β). The technical details of the BB are found in Khan & Morris (2014) and shown in Technical Appendix 5B. This approach uses a non-linear mixed modelling approach in which the general likelihood has been programmed directly using the SAS software (version 9.3). The SAS code for carrying out BB modelling of EQ-5D is found in Appendix 5A.2, with some technical details (Figures 5.2 and 5.3).

The BB regression model performed 'best' among all models. For the TOPICAL and SOCCAR trials, respectively, residual mean square error (RMSE) was 0.09 and 0.11; R^2 was 0.75 and 0.71; observed versus predicted means were 0.612 versus 0.608 and 0.750 versus 0.749. Mean

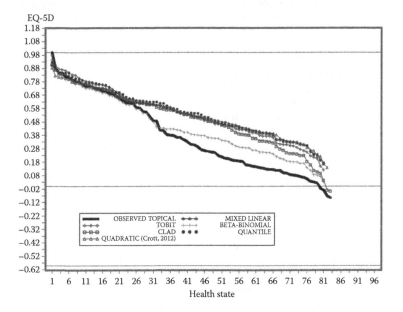

FIGURE 5.2
Predicted EQ-5D and observed EQ-5D for each model. Comparison of models using TOPICAL data. The x-axes are ordered health states (1 refers to 11111 and 101 is 23321); these are ordered according to the weighted value of the health state using the UK TTO tariff.

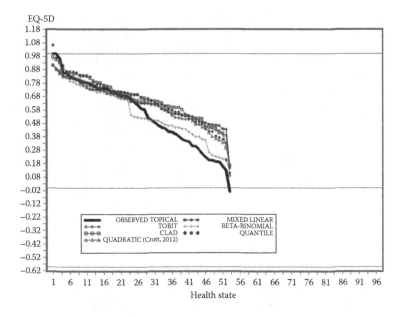

FIGURE 5.3

Predicted EQ-5D and observed EQ-5D for each model. Comparison of models using SOCCAR data. The x-axes are ordered health states (1 refers to 11111 and 63 is 23222); these are ordered according to the weighted value of the health state using the UK TTO tariff.

difference in QALYs (observed vs. predicted) was 0.051 versus 0.053 and 0.164 versus 0.162 for TOPICAL and SOCCAR, respectively. Models tested on independent data showed that the simulated 95% confidence limits from the BB model contained the observed mean more often (77% and 59% of the time for TOPICAL and SOCCAR, respectively) compared with the other models. All algorithms overpredict at poorer health states, but the BB model was relatively better, particularly for the SOCCAR data.

The BB regression approach showed superior performance compared with published models in terms of predicting the observed EQ-5D from QLQ-C30 in these lung cancer trials. This non-linear approach may offer advantages over existing models for mapping and as a general modelling procedure for utilities. This analysis also confirmed previous findings that mapping algorithms have been shown to overestimate the HRQoL at the poorer heath states. The BB regression model has shown some ability to predict these patients. The reasons why current algorithms persistently overpredict at poorer health states require further interrogation, perhaps incorporating adverse event information into the models using a single non-linear model which is flexible, offers results which have reasonable clinical interpretation, and avoids the need for arbitrary definitions of 'poor' and 'good' health states. The presence of differential treatment effects requires further research, particularly on how this impacts the predicted EQ-5D.

5.5 HRQoL Metrics for Use in Economic Evaluations

One key aspect of an economic evaluation is the derivation of the metric (or input) for generating the ICER and the cost-effectiveness acceptability curve (CEAC). There are two common metrics derived from raw HRQoL data used in an economic evaluation. These are

1. The QALY
2. The Q-TWiST

Both of these measures have relative advantages and disadvantages, which have been discussed extensively, particularly the QALY (Culyer et al., 2011). Not all may be accepted by every reimbursement authority, because some authorities (e.g. in Germany) prefer to consider the 'value package' rather than basing value on a single metric (IQwiG, 2008). Nevertheless, many local reimbursement authorities, such as NICE, consider the QALY as a key metric in their health technology assessments (HTAs) for demonstrating value.

5.5.1 QALY

QALY is essentially a measure of effectiveness that combines the quantity and quality of life into a single value. For example, a patient, after taking a new treatment that improved survival time to 10 months, might also have experienced HRQoL issues resulting in an average QoL, as measured by the EQ-5D utility, of 0.8. Combining both of these quantities ($0.8 \times 10 = 8$ months) gives a QALY value of 0.67 years (8/12). The derivation of a QALY can depend on several issues:

1. The assumption that the HRQoL data have been appropriately, accurately and reliably assessed and collected, so that key statistics such as the area under the curve (AUC) can be computed with a degree of precision. Assessing the EQ-5D at two time points only will not yield good estimates of the AUC.
2. Whether the QALY gain is a short-term or long-term gain. For example, where health benefits are expected beyond the follow-up period in a trial (e.g. longer survival), the benefits can be considered long-term, whereas in a trial that lasts only 12 weeks, the QALY gain is short term: once the treatment stops, HRQoL benefits disappear.
3. The modelling approach used to predict future utility or HRQoL values is adequate.
4. The nature and stage of the underlying disease. For example, if patients have exhausted all treatment options, such that

further survival benefits are unlikely, then a QALY may not be so important.

Several situations in which a QALY is calculated are now discussed, taking into account subtleties in calculations that are not always immediately apparent.

5.5.1.1 QALY Calculation Using the AUC with Patient-Level Data

A common metric for calculating a QALY is to compute the AUC. This example shows how the QALY is calculated for a single patient using EQ-5D utilities. The steps are as follows:

1. For each time point when the EQ-5D (or other preference measure) is assessed, convert the raw values into a utility index. Therefore, each patient will have an EQ-5D utility value at each time point. For example, if the EQ-5D is assessed every 3 months for 2 years, we might expect eight raw EQ-5D responses for each patient (although once a patient has disease progression, there may be fewer). The SAS code in Appendix 5A.1 can be used to convert the raw EQ-5D into utilities based on the TTO tariff.

2. If a given patient has lived for 18 months (died at 19 months), we might expect to have seven EQ-5D utility values: five in the first year (baseline, month 3, month 6, month 9, month 12) and two in the second year, just before death (month 15 and month 18). So, we are now in a position to compute the AUC over seven time points.

3. There is, however, one complication: the patient in this example died at 19 months, and the last EQ-5D assessment was at 18 months. The next assessment would have been at 21 months. Given that the patient was still alive between 18 months and 19 months, we need to decide on a value of the EQ-5D between 18 months and 21 months so that the EQ-5D can be computed. Clearly, the value at 19 months (and after) will be 0. We can also choose to ignore anything after 18 months and compute the AUC over the first 18 months: $AUC_{(0-18)}$. There is no right answer; it depends on beliefs about the disease and behaviour of the data. I will assume that after 18 months, the EQ-5D is 0 in this example, acknowledging that this might underestimate the AUC. This may not be unreasonable, because often disease progression precedes death, and the HRQoL is expected to be worse (even worse than death) around this time.

4. We now assume that the seven EQ-5D values at time 0 (baseline) and 3, 6, 9, 12, 15 and 18 months are 0.8, 0.7, 0.9, 0.8, 0.5, 0.3 and 0.2. The AUC using the trapezoidal rule is

$$[(0.8 + 0.7)/2 \times (3-0)] + [(0.9 + 0.7)/2 \times (6-3)]$$
$$+ [(0.8 + 0.9)/2 \times (9-6)] + [(0.5 + 0.8)/2 \times (12-9)]$$
$$+ [(0.3 + 0.5)/2 \times (15-12)] + [(0.2 + 0.3)/2 \times (18-15)]$$
$$= 11.1 \text{ months} = 0.925 \text{ QALY}$$

5. This process is then repeated for all patients in each treatment group, and all the AUC values are summarised to determine the mean QALY for each treatment group.

This calculation is simple, but assumes an absence of censoring. When there is no censoring, the sample mean of the QALYs could be calculated (for each arm) along with the standard errors. A standard two-sample t-test (for independent samples) can be used to compare QALYs. However, if censoring does exist, an alternative approach is needed.

One method to deal with censoring might be to multiply each patient's observed survival time by their mean utility (AUC or mean over time) to formulate a QALY end point (on a QALY scale). This is like doing a time to event analysis using the QALY as the analysis end point (because it is still in a time measurement, albeit weighted). However, the survival times of patients with poor HRQoL (low EQ-5D scores) will be down-weighted compared with those patients with a good HRQoL. For example, a patient with 6 month survival and a mean EQ-5D value of 0.2 gives a QALY of $6/12 \times 0.2 = 0.1$. Compare this with a patient who has a survival of 3 months, with a QALY = 0.2. The former patient will accumulate QALYs at a slower rate, and therefore will have a higher chance of being censored earlier (note that the x-axis of a Kaplan–Meier analysis will be QALY, not time). For a Kaplan–Meier analysis, one key assumption is that the survival times should be randomly censored that is, the censoring is not related to any factors associated with the survival time. Clearly, on a QALY scale, survival times are altered (by the QoL weights), resulting in a term known as *informative censoring*, which renders the standard Kaplan–Meier-type analyses invalid. One way to handle this problem is now discussed.

5.5.1.2 Calculation of the QALY Using Summarised EQ-5D Data When the Primary End Point Is Survival (Censored Data)

In Section 5.5.1.1, the QALY was determined using the AUC (or mean) at the patient level, and the mean QALY was just the simple arithmetic mean across all patients. In some trials, some patients may still remain alive, and the assessments for their HRQoL are considered incomplete. These patients are often censored. We therefore have a mix of patients, some who died and some who are still alive. A Kaplan–Meier curve is used to determine the proportion of patients alive up to a certain time point. Instead of computing the QALY using individual patient AUC values (or patient-level mean utilities), we can also compute the QALY by using the information from the survival curve and the EQ-5D data together (i.e. at the population level). However, the EQ-5Ds also need to be summarised. For example, we would first need to calculate the mean EQ-5D at each time point, and we would also need a value for the proportion of patients alive at the same time point. The two (summary) measures would then be combined to generate a QALY. This is different from calculating patient-level mean utilities and multiplying patient-level survival times to generate a QALY.

Although we restrict our estimates of the proportion alive and the mean EQ-5D to the duration of the trial, we can also estimate these values outside the trial follow-up period, that is, over the lifetime of patients. For example, if the trial lasted only 24 months, but some patients were still alive beyond 24 months, a model could be used to predict EQ-5D or the proportion alive after 24 months known as *extrapolation* (see Example 5.2).

The approach to computing a QALY with censored data is described in the following steps:

(i) Determine for each patient the EQ-5D index at each available time point. For example, there may be some patients who have values of EQ-5D observed to 12 months (died at 12 months) and others who have values to 24 months and even beyond (censored).

(ii) Specify a model that can be used to model the EQ-5D index over time. The purpose of the model might be to predict an EQ-5D value for each patient within each treatment group at each time point. Care should be taken that predictions are computed sensibly. It makes no sense to use a model to predict a patient's EQ-5D value at 18 months when the patient died at 12 months. The value of using the model is so that extrapolation of EQ-5D can be made for those patients still alive in the future, and also for predictions at other time points within the follow-up period.

(iii) Several (regression) model choices, such as adjusting for baseline, repeated measures and mixed effects models, can be used, as long as the mean utility can be estimated (and the model assumptions hold). It is not unreasonable to use a non-parametric model such as quantile regression, and use the mean of the predicted medians to adjust the survival proportions.

(iv) Once the mean EQ-5D at each time point is determined, the next step is to determine how to compute the proportion alive at each time point. The time points of interest are those at which the EQ-5D assessments were made, but as alluded to in step (iii), proportions at other time points could also be used, depending on the strength of the predictive model. The proportions alive can be determined using a Kaplan–Meier curve or a more sophisticated survival function (such as Weibull or cubic splines) for better fit.

Example 5.4: Calculating a Mean QALY Difference in the Presence of Censored Survival Data

In a clinical trial, the EQ-5D was assessed at baseline and every 2 months for the first year and every 3 months in the second year. All patients were followed up for a period of 2 years or until death. The expected QALY and motivation for determining the standard error of the QALY are provided for statistical inference around mean QALYs.

In this example, there are two end points: HRQoL and OS. In a full economic evaluation, the relationship between PFS and post-progression survival will also need to be considered.

In Table 5.2, Patient 1 progressed at 6.8 months, and upon further follow-up, died at 19 months post-randomisation. The last EQ-5D assessment was, however, available at 18 months after randomisation. Patient 2 neither progressed nor died and was censored at the last time of contact (24 months).

We start by first carrying out a standard survival analysis of the data set and use the Kaplan–Meier plots to generate the survival probabilities at each time point for the time to event end point (OS). The time points of interest are the same ones at which HRQoL measures are assessed. In addition, using a separate model, we also fit a mixed effects model of the following form, assuming that EQ-5D are normally distributed (ignoring censoring)

$$EQ\text{-}5D = Patient + Baseline + Treatment + Time + Treatment \times Time + error$$

This model assumes that the patient is a random effect. The within-subject effects are modelled (often termed *clustering*), and no specific correlation is imposed on the within-subject effects, although if the HRQoL is worsening over time, we could impose one. The SAS code for fitting such a model is

```
PROC MIXED DATA =<name>;
Class treatment patient time;
Model EQ5D = baseline treatment time treatment*time;
Repeated / subject=patient type=un;
Lsmeans treatment*time;
Run;
```

From this model, we can estimate the predicted mean EQ-5D for each time point for each treatment group. The TYPE = UN does not impose a correlation structure on the (within-subject) repeated measures. This model will provide predicted baseline adjusted mean EQ-5D for each patient at each time point. It is important that predicted estimates of EQ-5D make sense. For example, if a patient died at 4 months, the

TABLE 5.2

Example Data for Computing QALY When Censored Data Are Present (Example 5.4)

Patient	Treatment	Time Point	EQ-5D Index	PFS (months)	OS (months)	Details
1	A	Baseline	0.7			
1	A	2	0.7			
1	A	4	0.8			
1	A	6	0.8	6.8		Progressed
1	A	8	0.70			
1	A	10	0.6			
1	A	12	0.4			
1	A	15	0.3			
1	A	18	0.1		19	Died
2	A	Baseline	0.6			
2	A	2	0.5			
2	A	4	0.8			
2	A	6	0.9			
2	A	8	0.8			
2	A	10	0.8			
2	A	12	0.7			
2	A	15	0.75			
2	A	18	0.7			
2	A	21	0.7			
2 Etc.	A	24	0.6	24	24	Alive, progression free

predicted EQ-5D estimates at 4 months and all subsequent months should be equal to zero. A model could have been used to incorporate the censoring feature into QoL data, but this complicates the model considerably. The simple option is to set all EQ-5D values to 0 after death (which may also have implications for the correlations between EQ-5D over time).

For this example, the Kaplan–Meier survival times up to 2 years for Treatment A are as shown in Table 5.3 with predicted EQ-5D.

The expected (i.e. mean) QALY (or quality-adjusted survival [QAS]) is given by

$$QALY = \Sigma \left[EQ-5D_i + EQ - \frac{5D_{i+1}}{2} \right] \times \left[\frac{\left[S_i + \frac{S_{i+1}}{2} \right]}{\left[t_{i+1} - t_i \right]} \right] \quad (5.1)$$

TABLE 5.3

Kaplan–Meier Estimates and Predicted EQ-5D from a Regression Model

Time (months)	Survival S(t)	Predicted EQ-5D
0	1	0.63
2	0.9	0.72
4	0.85	0.66
6	0.70	0.57
8	0.65	0.42
10	0.53	0.45
12	0.49	0.36
15	0.37	0.47
18	0.31	0.37
21	0.27	0.29
24	0.15	0.27

$$= [\{0.63 + 0.72\} / 2 \times [1 + 0.9] / 2\} / (2 - 0)$$

$$+ \{0.72 + 0.66] / 2 \times [0.9 + 0.85] / 2\} / (4 - 2) + \cdots]$$

This calculation would then be repeated for the other group in order to calculate the mean QALY difference.

In order to compute the standard error, bootstrapping or simulation can be used. To compute the standard error using bootstrapping, the program provided in Chapter 3 can be used by first generating 1000 bootstrap samples of the same sample size as the trial (if the clinical trial had 300 patients, then this would be 1000 bootstrap samples, each of size 300) with varying survival times, and, separately, utilities for each of the 1000 bootstrap samples. From each of these bootstrap samples, the survival rates can be computed at each time point along with the mean utilities. Then the formulae in Equation 5.1 can be used to compute the mean, variance and standard errors of the expected QALYs. The purpose of the standard errors is for statistical inference around mean difference in QALYs. For PSA, the 10,000 QALYs can be generated using Monte-Carlo simulation.

5.5.1.3 Adjusting for Pre-Progression and Post-Progression HRQoL

One important aspect of HRQoL, particularly in cancer trials, is that the HRQoL can change rapidly after the disease has progressed. Therefore, mean utilities at each time point should take into account the pre- and post-progression responses. One approach is to consider the overall survival as consisting of two components, PFS plus post-progression survival (PPS): OS = PFS + PPS. Hence the mean EQ-5D can be computed during the pre- and post-progression periods by multiplying each survival component by the pre- and post-progression EQ-5D

$$\text{Expected QALY} = \text{Pr ogression} - \text{Free EQ-5D} \times \text{Mean PFS}$$

$$+ \text{Post-progression EQ-5D} \times \text{Mean PPS}$$

Example 5.5: Computing the QALY Using Mean PFS and Mean PPS

The mean OS was 6 months, of which the PFS was 4 months. The mean EQ-5D during the PFS period was estimated by calculating the mean EQ-5D during each of the pre- and post-progression periods; hence, the mean pre-progression EQ-5D was 0.7, and the mean EQ-5D after progression was 0.4. The PPS is $6 - 4$ months = 2 months. Hence, the mean QALY is $0.7 \times 4 + 0.4 \times 2 = 2.8 + 0.8 = 0.3$ QALYs.

Note that this method does not weight the survival times by mean EQ-5D as in Example 5.4. If EQ-5D estimates at each time point are required that take into account pre- and post-progression EQ-5D at each time point, a model-based approach can be used.

Example 5.6: Computing the QALY Using Mean PFS and Mean PPS from Model-Based Estimates

Patients from an RCT were assessed for EQ-5D at baseline and twice monthly for 2 years (a total of 12 measurements). The Kaplan–Meier estimates for the experimental treatment at each 2 month time point (2, 4, 6, 8, 10, 12, 14, 16, 18, 20, 22 and 24) are 90%, 85%, 77%, 65%, 60%, 56%, 43%, 38%, 35%, 26%, 14% and 8%.

A mixed effects model was used to estimate the mean EQ-5D at the same time points while taking into account whether the response was before or after disease progression. Hence, in the data set, an indicator variable was set up such that if an EQ-5D measurement (at any time point) was prior to the patient progressing, it was coded as 0; otherwise, it was coded as 1. The mixed model was of the form

$$\text{EQ-5D} = \text{Subject} + \text{Treatment} + \text{Pr ogress Time} + \text{Time} + \text{error}$$

where Progress Time is the indicator variable for when the patient progressed.

From this model, the mean EQ-5D can be estimated for each treatment group, at each time point, after adjustment for whether the measurement was prior to or after progression. The SAS code for generating the mean EQ-5D estimates at each 2 month time point is

```
PROC MIXED DATA =<name>;
Class treatment patient time prepost;
Model EQ5D = baseline treatment time prepost;
Repeated / subject=patient type=un;
Lsmeans treatment time;
Run;
```

The variable *prepost* in this code is an indicator variable for whether the response was prior to or after disease progression.

The estimates of mean EQ-5D from each of the time points can then be estimated and applied using Equation 5.1 to estimate the mean QALY. Bootstrapping can be used to estimate the standard error as described earlier.

5.5.1.4 Calculation of the QALY Using Summarised EQ-5D Data When the Primary End Point Is Not a Survival/ Time-to-Event End Point (Short-Term Health Gains)

When the primary end point is survival time, the approaches described in Section 5.5.1.2 for estimating the mean QALY are adequate. When the end point is not survival time, there are some issues that need to be considered when deriving the QALY.

(i) Differences between treatments are expected to exist at one or more time points during the follow-up period (e.g. at the 12 week assessment for a primary end point defined as the change from baseline in pain score at 12 weeks). The treatment differences become negligible at later time points, particularly after treatment has stopped. Therefore, estimation of the mean QALY is computed over the duration of the key follow-up period only. For example, in a 24 week study of pain, the primary end point might be at 24 weeks (for reaching an optimal treatment benefit). After 24 weeks, the treatment effects start to disappear. The time horizon for an economic evaluation might also be over the first 24 weeks, with the assumption that QALY gains are zero thereafter. If (as happens in some clinical trials) an additional open label period from week 24 to 52 weeks with maintenance therapy occurs, extending the time horizon for an economic evaluation over 1 year is also possible. In some clinical trials, however, such as for diabetes, the short-term end points do have implications for longer-term health benefits, and it is not uncommon to have a primary end point (such as reduction in glucose after 4 weeks) with future health benefits expected several years into the future.

(ii) Some health conditions (e.g. back pain) do not impact patients' survival time, but do impact their HRQoL. New treatments should, therefore, not only demonstrate value for the primary efficacy end points but also show either value for several other secondary end points or, at least, no loss in value from them (i.e. the new treatment does not show worse effects on average).

(iii) Following on from (i), in order to estimate the short-term QALY gain more accurately, there should be a sufficient number of HRQoL assessments to estimate the AUC with some degree of precision. For example, having only two valid HRQoL assessments (month 1 and month 2 assessments) from a 12 month trial is unlikely to yield a reliable estimate of the AUC. It is not uncommon for clinical trials with short durations to be designed for detecting treatment effects at specific time points. For example, powering a study at 12 weeks and taking many assessments close to baseline and 12 weeks with few measurements between baseline and 12 weeks will not yield a very good estimate of the short-term QALY gain.

(iv) No Kaplan–Meier curves or survival analysis methods are employed, and therefore no issues of censoring arise (although missing data and dropouts might still be an issue). Health benefits are often not considered to exist outside the trial follow-up period, except in some disease areas, such as diabetes or arthritis, for which the primary clinical end point is relatively short term (e.g. at 4 weeks), but the implications for future health benefits can be extrapolated up to several years outside the clinical trial follow-up period.

Therefore, if we wish to perform a CUA, but the objective is not concerned with life years gained, we need an alternative approach to computing the QALY.

Example 5.7: Computing a Short-Term QALY Gain

In a 12 week randomised double-blind trial for pain (e.g. Dunlop et al., 2012), EQ-5D values were obtained at baseline and at weeks 2, 4, 8 and 12. The QALY was determined up to 43 weeks (i.e. the expected duration of the treatment) by simply computing the mean or AUC for each patient. The five observed EQ-5D values over the 12 weeks were plotted, and values after week 12 were assumed to remain constant (i.e. constant from week 12 to week 43). The QALY is the AUC (Figure 5.4).

The mean EQ-5D utility score for all patients at baseline was 0.6, and for each of the four post-baseline time points (weeks 2, 4, 8 and 12) it was 0.70, 0.55, 0.59 and 0.65, respectively. This simple QALY has the uncertain assumption that the utility is constant between weeks 12 and 43, which may not be realistic.

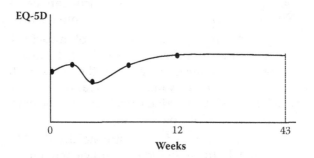

FIGURE 5.4
AUC assuming that the QoL is constant after week 12.

**Example 5.8: Computing a Short-Term QALY Gain:
Markov Model Set-Up**

In this example, the primary end point is from a rheumatoid arthritis trial. The trial was conducted as a double-blind trial with the primary end point of a reduction in morning stiffness (MS) (Dunlop et al., 2013). In this economic evaluation, the QALY was calculated at the population level – that is, the QALY was generated after efficacy and HRQoL data had been summarised.

The primary end point in the trial was defined as a change from baseline in morning stiffness at week 12. This was derived as a mean score. Since a Markov model (Chapter 6) was employed, the primary end point needed to be classified into durations of MS. The classifications are called health states. Table 5.4 shows the probabilities of patients falling into each of the four health states, for each of the treatment groups. Hence, this is called a four state Markov model. Obviously, careful thought needs to be given as to why and how these health states were defined.

In Table 5.4, the proportion of patients who fell into each of the K health states at each time point ($t = 0–3$) was taken from the clinical study report (CSR). The sum for each row is 1. For example, at baseline, about 10% experienced mean MS of <1 h. A useful treatment would expect to increase these proportions over time. We should also expect to observe patients moving out of the poorer health states (e.g. MS 2–3 h) into the better health states over time.

In order to compute the QALY, we also need to know the utility (EQ-5D) associated with each health state. This is shown in Table 5.5.

The values U_1–U_4 are the mean utility values computed across all patients within each health state. In this example, we assume the utilities are the same for each treatment group. Hence, the QALY will be determined by multiplying the proportion of patients who fall into each of the health states by the mean EQ-5D scores in Table 5.5

$$Q_j = \sum_{t=0}^{t11} \left(P_{jt1} \times U_1 + P_{jt2} \times U_2 + P_{jt3} \times U_3 + P_{jt4} \times U_4 \right)$$

where:

Q_j = total QALY for treatment j
K = health states 1–4 (i.e. U_1, U_2, U_3, U_4 refer to K = 1, 2, 3, 4)
t = time
P_{jtk} = proportion of patients in each health state for treatment j across health state k at time t
U_k = the utility values for each health state

Hence, the QALY for the MR group is calculated as ($0.095 \times 0.78 + 0.276 \times 0.61 + 0.240 \times 0.5 + 0.389 \times 0.45$) + ($0.341 \times 0.78 + 0.261 \times 0.61 + 0.120 \times 0.5 + 0.278 \times 0.45$) + $\cdots$ = 0.62.

TABLE 5.4

Morning Stiffness Classified into Four (K = 1–4) Health States

Time Point	MR-prednisone (%)				IR-prednisone (%)			
	MS <1 hour $K = 1$	MS 1–2 hours $K = 2$	MS 2–3 hours $K = 3$	MS ≥3 hours $K = 4$	MS <1 hour $K = 1$	MS 1–2 hours $K = 2$	MS 2–3 hours $K = 3$	MS ≥3 hours $K = 4$
Baseline ($t = 0$)	0.095	0.276	0.240	0.389	0.095	0.276	0.240	0.389
Month 1 ($t = 1$)	0.341	0.261	0.120	0.278	0.213	0.207	0.179	0.401
Month 2 ($t = 2$)	0.391	0.273	0.114	0.221	0.268	0.210	0.181	0.341
Month 3 ($t = 3$)	0.432	0.223	0.108	0.237	0.262	0.264	0.126	0.348

Note: MR: modified release; IR: immediate release.

TABLE 5.5

EQ-5D Utility Scores

Health State (K)	Mean Utility Score (Uk)
MS <1 h (U_1)	0.78
MS ≥ 1 and <2 h (U_2)	0.61
MS ≥ 2 and <3 h (U_3)	0.50
MS ≥ 3 h (U_4)	0.45

After week 12, the patients were assumed to remain in the same health state, and therefore the QALY gain was 0 (i.e. the QALY remains at 0.62 after 12 weeks in this example for the MR group). The process is repeated for the IR group so that the mean QALY difference can be computed.

5.5.2 Q-TWiST

Q-TWiST is another type of QALY, which combines the survival time and the quality of life experienced during this period into a single measure. There are two components to the Q-TWiST:

(i) The TWiST component
(ii) The QoL component (utility) associated with (i)

5.5.2.1 The TWiST Component

In (i), the amount of time without symptoms of the disease or toxicity is computed as follows:

(a) Compute the time from randomisation to disease progression (or relapse) (PFS).
(b) Within this period of time, compute the periods of time the patient experienced toxicity (there can be more than one). This is typically from the date of documented start of toxicity (e.g. neutropenia grade 3 and above) to the date of resolution or stopping. These are assessed prior to relapse or progression.
(c) Compute the periods of time for which the patient did not have symptoms of the disease or toxicity.
(d) The TWiST component is (a) minus (b).

Figure 5.5 shows how the TWiST component is calculated: by subtracting the grey parts (periods with toxicity) from the PFS time.

Since the TWiST is still a time to event end point, comparing treatments using standard Kaplan–Meier methods can result in biased estimates due to informative censoring: a patient with a longer duration of toxicity is likely to

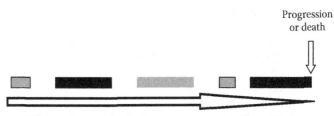

FIGURE 5.5
Graphical description of the TWiST component: TWiST is the PFS time minus the sum of the grey parts (toxicity periods).

be censored before a patient with a shorter duration. One approach is to set a cut-off limit L for the amount of TWiST – for example, all patients should have a TWiST for at least 6 months (L = 6). If all patients have TWiST >6 months, comparisons between mean TWiST can be made by computing the area under the TWiST survival curve using Kaplan–Meier methods. However, even with low censoring (Gelber & Goldhirsch, 1989), this approach (using the cut-off L) can give biased estimates (bias in this context refers to comparing mean values without censored data with data which may be censored). Alternative approaches are found in Gelber (1989).

5.5.2.2 Adding HRQoL with the TWiST Component

One way of incorporating utility into the TWiST is to say that if a patient experiences toxicity during the follow-up period, the EQ-5D is set to 0; otherwise, it is set to 1. The Q-TWiST is then a simple weighted (by 1 or 0) sum of the survival times. An alternative approach would be to weight the times by values other than 0 or 1 (e.g. a value between 0 and 0.5 if toxicity exists and between 0.5 and 1 when toxicity does not occur). The weights would need careful justification, otherwise the mean Q-TWiST differences could be biased (e.g. the weights could be chosen with preconceived ideas that the new treatment offers value). A more involved approach is to define health states and attach weights to each health state. The survival time can then be partitioned so that the mean amount of time spent in each health state can be calculated. The sum of the parts would then form the basis of a QALY based on Q-TWiST. Note, however, that in many cancer trials, there is unlikely to be a state in which patients have no signs or symptoms; therefore, careful thought needs to be given to how heath states are defined.

Example 5.9: How to Derive a QALY from Q-TWiST

In this simplified example, we assume the disease consists of pain, and we wish to compare two treatments, A and B. We define two health states, H_1 and H_2, where H_1 is a state of constipation, a known gastrointestinal side effect of opioids, and H_2 is no constipation, that is, the TWiST state. The objective is to compute the amount of time each patient has in either health state. Utilities U_1 and U_2 from observed EQ-5D are then computed for patients within each of the health states. If a patient experienced both states (i.e. improved constipation after initially being worse), the mean utility for each health state should reflect this. The QALY using Q-TWiST is then computed as

$$Q\text{-}TWiST = U_1 \times h_1 + U_2 \times h_2$$

where h_1 and h_2 are the mean times spent in each of the (true) health states H_1 and H_2, respectively. The calculation differs from the oncology setting because there is no survival time (only follow-up time). The total follow-up time for each patient is used, and the periods of time in which constipation, as defined by the bowel function index (BFI), was >28.2 can be weighted by the utility in that state and subtracted from the follow-up time. For example, if a single patient was followed up for 24 weeks, during which the BFI was <28.8 (no constipation) between weeks 0 and 20 and >28.8 in the last 4 weeks, the TWiST would be $24 - 4 = 20$ weeks. However, each component part needs to be weighted by the mean utilities

 (i) Between weeks 0 and 20 (no constipation) (state H_2)
 (ii) Between weeks 20 and 24 (constipation) (state H_1)

Clearly, the utility is expected to be higher in (i) than in (ii). Hence, the Q-TWiST for this patient is

$$U_1 \times 20 + U_2 \times 4$$

If the patients had U_1 and U_2 calculated as 0.7 (no constipation) and 0.4, respectively, the Q-TWiST would be $0.7 \times 20 + 0.4 \times 4 = 14 + 1.6 = 1$ 5.6 weeks. That is, after partitioning the follow-up time into two health states, the mean time without signs or symptoms of constipation was 15.6 weeks. For many patients, these values could be summarised by treatment group. In this example, we assumed only two health states. For many health states, the calculations become tedious. Determination of the utilities, assuming patient-level HRQoL data are available, requires all patients to be classified into their respective health states, and the mean utility within each health state needs to be derived.

5.5.3 Disability-Adjusted Life Year (DALY)

Another measure used in economic evaluation (but not as common as the QALY) is the DALY. The DALY is essentially the inverse of the QALY. We think of the QALY gained, and its opposite is the DALY averted. An example of using DALYs might be to compare policies for safe and appropriate use of injections, because poor injection practices can result in potentially life-threatening situations (Dziekian et al., 2003).

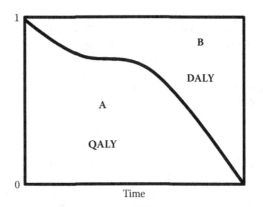

FIGURE 5.6
Relationship between QALY and DALY.

In Figure 5.6, the QALY is the AUC (part A) and part B is the amount of DALY 'lost'. What this means is that a patient who survives for 6 months with a utility value of 0.7 has essentially lost utility of 0.3. If the HRQoL were perfect, 6 months of survival improvement would be valued at 6 months. The QALY values this at $0.7 \times 6 = 4.2$ months, so 1.8 months of QALY lost is the value of the DALY. Although economic evaluations assess cost-effectiveness in terms of the cost/QALY, many decisions cannot be fully explained by the cost/QALY (see Culyer, 2011 for further details).

5.5.4 Statistical Issues with HRQoL Data

5.5.4.1 Missing QoL Data

One common feature, as noted earlier, is the presence of missing HRQoL data, particularly missing utilities. The raw EQ-5D scores are converted into utilities based on the use of a TTO tariff (Dolan, 1990). For example, if the raw EQ-5D scores for a particular patient are 21321 for each of the five domains, this corresponds to a utility of 0.364. However, it is quite possible (and likely) that a patient does not have all five responses. What happens if they have three of the five, resulting in the health state '223xy' (where x and y correspond to missing values)? The TTO algorithms do not provide a utility for partial health states other than for this particular configuration of responses. Imputation methods may result in values that are non-integers, and median or mode imputation may not be suitable, because a median may still result in a decimal value, and there may be two modes. Clearly, setting a patient's utility to *missing* with responses such as 2233y (last value missing) is not satisfactory, because information (20% of the data) is lost. It is unclear from literature how these issues should be handled; Simons et al. (2015) compare various imputation approaches, but there still appears to be a lack of guidance on whether to impute at the domain level or at the utility index value level. One approach might be to use an imputation approach (or model) to simply estimate the missing utility, but this does not make use of the observed EQ-5D data available. Another approach might be to use the maximum of non-unique median imputations.

More complex issues of analysing HRQoL data with missing data are not discussed further. Several excellent references can be found in the bibliography, among which Carpenter & Kenward (2013) and Fairclough (2010) cover the mechanisms of missingness, such as missing at random (MAR), missing not at random (MNAR) and missing completely at random (MCAR).

5.5.4.2 Effect Sizes

Often, the clinical effect sizes for HRQoL end points are not known. For EQ-5D (as for the SF-36 and HUI), the clinically relevant effect sizes are almost always unknown (this author has not come across any EQ-5D effect size that has been suggested to be of clinical relevance). One reason for this is that EQ-5D is not a measure of clinical efficacy. It is an adjustment to the primary (and sometimes secondary) end points for the purposes of an economic evaluation. Therefore, to analyse EQ-5D with a view to establishing a statistically significant difference between treatments may be unnecessary and misleading, although for 'marketing purposes' it might add some 'dazzle' to the results. For other generic measures, too, unless some precedent is set, the magnitude of a clinically relevant HRQoL difference contains much uncertainty. Some have suggested, as in the case of the QLQ-C30, effect sizes of at least 10 points (Osoba, 1998). Cohen (1998) categorises effect *sizes* (Cohen's d) as small, medium or large (small: 0.2–0.3; medium: about 0.5; large: above 0.8). However, since the effect size is the treatment difference divided by the standard deviation, it can be difficult to interpret these measures from a clinical perspective. For the purposes of value arguments, making a simple statement such as 'a 10 mmHg difference (reduction) with a new treatment compared with the standard' is clear and transparent. Stating this reduction as an effect size of, say, 3.4 (10/2.9: mean/SD) makes it difficult to interpret clinically, even if this effect size is categorised as 'medium'. One may also consider alternative measures of effect, such as odds ratios, which may be more meaningful to both patients and clinicians (Khan, in press).

5.5.4.3 Methods of Analysis

Most methods of analysis of HRQoL use some form of repeated measures analysis because the responses are collected on more than one occasion. There are several useful references for analysing HRQoL data (e.g. see Fairclough, 2010). The most common methods associated with analysis of the EQ-5D utilities, including those mentioned in Section 5.4.1, are

(i) *Repeated measures models,* assuming EQ-5D are normally distributed. This is a common approach, although for PSA, the distribution of EQ-5D has sometimes been assumed to follow a BB distribution. The advantage of model-based methods is that mean utilities can be adjusted for several prognostic factors and imbalances in the data structure.

(ii) *Mixed effects models*, which extend the techniques in (i) and use subject as a random effect. The advantage of mixed models lies in their ability to handle within-subject effects and repeated measures, and they may be efficient in handling some types of missing data.

(iii) *Methods that take censoring into account*, such as TOBIT and CLAD models (Section 5.3). However, models such as TOBIT and CLAD are still limited because censoring does not occur in relation to time, but in relation to specific responses: for example, if negative values (EQ-5D <0) are observed for EQ-5D, these could be restricted such that they equal 0.

Whatever models are used, justification for their choice and model diagnostics should always be checked.

5.6 Are Utility Measures Sensitive Enough for Detecting Treatment Differences?

Although the EQ-5D-3L is considered to be an important generic measure of HRQoL used in economic evaluation, concern has been expressed that such generic HRQoL measures are not as sensitive to detecting treatment benefits as some of the (longer) disease-specific cancer instruments such as the EORTC-QLQC-30. Consequently, Khan (in press) compared the treatment effects of the QLQC-30 and EQ-5D, and found significant underestimation of treatment differences based on the odds ratio and mean difference scale. Further details can be found in Khan (in press). However, the main point of the analysis in Khan (in press) is that adjustment to the EQ-5D utilities should be made to compensate for the loss of HRQoL that would otherwise have been gained if a more sensitive measure had been used. This is an interesting and timely analysis, particularly since it is increasingly difficult to show differences between treatments after EQ-5D adjustments to primary outcomes in the context of economic evaluation.

Exercises for Chapter 5

1. Which HRQoL instruments would you use when planning an economic evaluation in lung cancer, and why? What concerns would you have about the EQ-5D-3L?

2. When EQ-5D utilities are not available or collected in a clinical trial, how might you estimate them?

3. How would you go about developing and testing a mapping algorithm?

4. Discuss the relative advantages of the QALY, Q-TWiST and DALY as potential measures of demonstrating the value of a new treatment.

5. Explain why you might use a model-based estimate of EQ-5D utilities as opposed to those based on raw summary statistics.

6. Describe how the QALY is calculated in a cancer trial in which EQ-5D utility data are collected every month for 24 months.

7. How might it be possible to compare the sensitivity of a generic measure with a condition-specific measure?

8. What is the difference between the EQ-5D-3L and the EQ-5D-5L? Discuss with your colleagues how you might show which is more sensitive to showing differences between two or more healthcare technologies.

Appendix 5A: SAS/STATA Code

5A.1 SAS Code for Converting Raw EQ-5D-3L Responses into Utilities

```
data eqg;
set eq;**get raw response eq5d data**;
equk=1;
if (eqmob = 2) then equk = equk - 0.069;
if (eqmob = 3) then equk = equk - 0.314;
if (eqsc = 2) then equk = equk - 0.104;
if (eqsc = 3) then equk = equk - 0.214;
if (equa = 2) then equk = equk - 0.036;
if (equa = 3) then equk = equk - 0.094;
if (eqpa = 2) then equk = equk - 0.123;
if (eqpa = 3) then equk = equk - 0.386;
if (eqdep = 2) then equk = equk - 0.071;
if (eqdep = 3) then equk = equk - 0.236;
if (eqmob ne 1) or (equa ne 1) or (eqsc ne 1) or (eqpa ne 1)
or (eqdep ne 1) then equk = equk - 0.081;
if (eqmob = 3) or (eqsc = 3) or (equa = 3) or (eqpa = 3) or
(eqdep = 3) then equk = equk - 0.269;
if (eqmob = .) or (eqsc = .) or (equa = .) or (eqpa = .) or
(eqdep = .) then equk = . ;
output;
run;
```

5A.2 SAS/STATA Code for BB Regression (Mapping Algorithm)

In STATA the command is trivial:

```
betareg EQUK SFbas SFpbas
```

where EQUK is the EQ-5D index (transformed on a scale of 0 to 1), SFbas is the baseline social function score and SFpbas is the post-baseline SF score from the EORTC-QLQC-30.

IN SAS, the SAS code is much more complex.

```
proc nlmixed data=x method=ml tech = hess cov itdetails;;
parms b0=2 b1=4;**initial estimates intercept and regressor
coefficient**;
*linear predictors;

Xb = b0+ b1*sf2; **modelling EQ-5D as a function of SF from
QLQC-30**
Wd = d0+ d1*sf2; **variance of EQ-5D as a function of SF from
QLQC-30**

*link functions transform linear predictors;
mu = exp(Xb)/(1 + exp(Xb)); ** this is the link function for
the mean**;
phi = exp(-Wd); ** and for the variance**;
*transform to standard parameterization for easy entry to
log-likelihood;
w = mu*phi;
t = (1-mu)*phi;

*log-likelihood;
ll = lgamma(w+t) - lgamma(w) - lgamma(t) + (w-1)*log(z) +
(t-1)*log(1-z);
model z ~ general(ll);

predict mu out = equk_pred_hat;**predicted values of EQ-5D**;
run;
```

Technical Appendix 5B: Beta Binomial Technical Details

The BB has the function

$$F(y \mid \alpha, \beta) = \left\{ \frac{\Gamma(\alpha + \beta)}{\Gamma(\alpha)\Gamma(\beta)} \right\} \times y^{\alpha-1} \times (1-y)^{\beta-1}$$

This form of the model is not useful for modelling and requires re-parameterisation so that the mean and variance can be modelled as functions of the predictors. The mean and variance of a beta (α,β) function are $\alpha/(\alpha+\beta)$ and $\alpha\beta/(\alpha+\beta)^2(\alpha+\beta+1)$, where α and β are the shape and scale parameters, respectively. Previously, the BB regression approach has been used in an attempt to model the shape parameters and not the simpler transformation of the shape parameters in terms of the mean and precision.

By specifying two link functions, from the mean and precision parameters – $h(\mu) = \text{logit}(\mu)$ for the mean; $g(\phi) = \log(\phi)$ for the precision the general likelihood in terms of the QLQ-C30 predictors can be estimated from the likelihood function

$$L(\beta,\delta,Y,X,W) = \frac{\Gamma(\exp(W\delta))}{\Gamma(s)\Gamma(t)Ys-1(1-Y)t-1}$$

$$s = \frac{\exp(X\beta + W\delta)}{\{1+\exp(X\beta)\}}$$

$$t = \frac{\exp(W\beta)}{\{1+\exp(X\delta)\}}$$

where X and W are a matrix of independent variables and β and δ are coefficient estimators for the mean and precision parameters, respectively. This form of parameterisation allows a powerful way of modelling utilities not only for mapping, but in a generalised mixed modelling context for estimates of utility increments.

One reasonable assumption we impose using BB is that EQ-5D is measured on a scale of 0–1, and observed values <0 are either ignored or set equal to 0. We chose to set observed EQ-5D <0 to 0 to avoid losing data. There were <1% of values with EQ-5D responses <0 in each data set; therefore, the potential for bias is likely to be small.

Predictions of EQ-5D are slightly complicated and involve inverting the usual logit function. For example, when applying the algorithm in practice, for given coefficients $\beta_1...\beta_{15}$, the predicted EQ-5D is estimated as

$$\text{Log}\left[\frac{\pi}{(1-\pi)}\right] = \alpha + WX$$

$$\text{So that } \pi = \frac{\exp\{(\alpha+WX)}{[1+\exp(\alpha+WX)]}$$

where:

   π contains the true EQ-5D response

   W contains the coefficients

   X contains the QLQ-C30 responses

Technical Appendix 5C: Technical Summary of the GLM

Let the outcome Y (costs) be related to a set of predictor variables. The linear relationship is specified as

$$\text{If } Y = XB + \varepsilon$$

And therefore, the expected value of Y, $E[Y] = XB$ (because $E[\varepsilon] = 0$).

Since we wish to estimate the population mean, μ, $E[Y] = \mu$, the GLM generalises linear regression using several important concepts:

(i) A link function: for example $g(XB)$

(ii) A linear predictor (e.g. treatment and some other covariates)

(iii) The inverse link function: for example $g^{-1}(XB)$, which converts (ii) to a mean

(iv) A relationship between the mean and variance

In the situation where Y has a normal distribution, the GLM requires

(a) The link function, $g(Y) = g(XB) = 1$.

(b) The linear predictor is $\eta = \beta_0 + \beta_1{}^*\text{Treatment}$, where the mean costs is assumed to depend on treatment only.

(c) The inverse link function is η.

(d) The relationship between the variance of $E[Y] = XB = \mu$ is considered to be constant; hence $\text{Var}(Y) = \text{Var}\{g^{-1}(XB)\} = \text{Var}\{g^{-1}(\eta)\} = \text{Var}(\mu) = 1$.

In this case, the OLS and GLM estimates will be the same because the link function is linear. In non-linear link functions (e.g. taking logarithm of costs), estimates from GLM and OLS models may not be the same.

6

Modelling in Economic Evaluation

6.1 Introduction to Modelling: Statistical versus Economic Modelling

In Section 2.5, we introduced different concepts of modelling, distinguishing between statistical models where patient-level data are available and decision models that use published or summarised data from clinical trials. If patient-level data on costs and effects exist, then using available data is preferred to modelling. However, patient-level data may not be available (due to limited follow-up time), and therefore an alternative approach to estimating the mean costs, effects and incremental cost-effectiveness ratio (ICER) is required. One way is to use published data or data summarised in the clinical study report to build the economic model. In this chapter, we describe how to use data from summarised clinical reports or published data as well as patient-level data to estimate the ICER. In Chapter 7, we will learn how to quantify the uncertainty around the ICER in sensitivity analyses.

The merits of decision modelling have been discussed elsewhere (Gray et al., 2011); however, it may not be clear to a researcher whether one should carry out a stochastic analysis using the patient-level data or just summarise the data and build a decision analytical model. For example, in a clinical trial of two treatment groups (Treatments A and B) which has also been designed for a prospective cost-effectiveness analysis, a patient-level approach might be suitable. However, if summarised data on a third treatment (Treatment C) was available from a separate clinical trial, a decision model approach could be used, which might allow a comparison of effects and costs between all three treatment groups. A full patient-level analysis might not be possible, because patient-level data are not available for treatment group C. Therefore, an alternative approach might be a decision analytic–type model (or even some form of evidence synthesis, discussed in Chapter 9).

In this chapter, we will learn how to construct a model for cost-effectiveness using three common approaches:

(i) Decision tree approaches
(ii) Markov models using summarised clinical trial data (or other data sources) which have previously been summarised
(iii) Stochastic or patient-level analysis

6.2 Decision Tree Models

A typical decision tree model structure is shown in Figure 6.1. The 'tree' has branches that relate to either actions or decisions taken by patients or clinicians, or outcomes (such as measures of treatment effect). Square nodes denote a decision between alternative treatments (decision nodes), and circular nodes indicate events (pain/no pain). Each branch has a probability (expressed as a rate, percentage or proportion) and an associated cost. The probability (or percentage) could relate to a decision, or it may also relate to a measure of effect (e.g. 20% had pain relief or 30% were given other treatment options).

For each branch, the expected costs and effects are calculated by multiplying the probability of decisions/outcomes by the costs for the branch. The expected costs are then totalled for each treatment and used to derive the ICER. The most important part, as in all models, is to determine the inputs (i.e. justification for

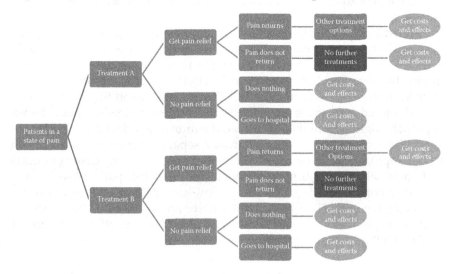

FIGURE 6.1
Simple decision tree structure for treating pain.

the choice of probabilities, rates, costs, etc.). The important model inputs are obtained from either published clinical trials, observational studies or surveys.

An important aspect of designing the 'branches' of the tree is understanding the treatment pathway, that is, what happens to patients in terms of their treatment plan and how they are likely to be followed during their treatment programme. For example, in Figure 6.1, the pathway might be

1. Patient starts taking one of two treatments to treat the disease (A or B).
2. Patient experiences some pain relief.
3. If the patient experiences pain relief, does the pain return? (Note: we are not asking how long pain relief lasts, but we could do so by forming a separate branch.)
4. If there is no pain relief, what happens next? Does the patient simply take no further treatment and have a reduced quality of life (utility) or, even worse, need to visit the hospital or stay in hospital?

Decision trees are useful when outcomes are likely to occur soon. Infections, chronic diseases, death events which might occur quickly, and other outcomes which occur over a short time window are suitable for decision tree problems. In a disease such as cancer, if branches of the tree were 'Disease Progression' or 'Response', it would be possible for a patient to have a quick 'Response' followed by disease progression 2 years later, and then die shortly afterwards. Using a decision tree might require two branches: one for 'Progression before 6 months' and the other for 'Progression after 6 months'. However, costs and effects accrue by the month, and in particular, the chance of progression before 6 months will depend on what the chance of progression was at 5 months. In this sense, time dependency would not be taken into account using a simple decision tree. Several branches might be needed (for each month), which can become difficult to interpret.

Example 6.1: Decision Tree Used to Estimate the Cost/QALY for a Clinical Trial in Pain

Data are presented in a summarised form (Figure 6.2) from a (fictitious) clinical trial with a form of the decision tree structure slightly modified from that of Figure 6.1. There were two treatments (A and B). Pain was the main clinical outcome reported in the trial, along with utilities. Resource use was not collected, and therefore costs had to be estimated from published literature. Utilities were estimated from health-related quality of life (HRQoL) collected during the trial, but these were condition-specific measures. Therefore, a mapping function (Chapter 5) had to be used to calculate utilities.

In Figure 6.2, the chance of a patient having pain relief on Treatment A is 0.7, and among those who did get pain relief, the chance of the

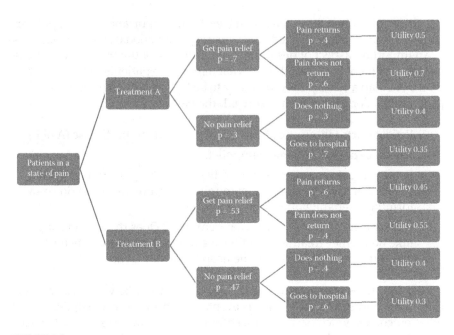

FIGURE 6.2
Decision tree for Example 6.1.

pain returning was 0.4. There are four treatment pathways for each of Treatments A and B (eight branches in total). The costs and effects (utilities) need to be calculated for each pathway. Table 6.1 gives a summary of the cost and utility components for each pathway with expected costs and effects.

TABLE 6.1

Computation of Expected Costs and Effects for Example 6.1

		Treatment A			
Path[a]	Probability	Utility	Costs (£)	Effect	Expected Costs (£)
PR/RT	0.28[b]	0.50	2000	0.14[c]	560[d]
PR/NNRT	0.42	0.70	3200	0.294	1344
NPR/N	0.09	0.40	1800	0.036	162
NPR/H	0.21	0.35	1500	0.074	315
Total	**1.00**			**0.544**	**2381**

[a] NPR/H: no pain relief followed by hospitalisation; NPR/N: no pain relief followed by no further treatment; PR/NNRT: pain relief without a return of pain; PR/RT: pain relief followed by returning pain.

[b] This probability is computed by multiplying 0.7 by 0.4 or Pr(A) × Pr(B|A), where Pr(A) is the probability of pain relief and Pr(B|A) is the probability that pain returns given that there is pain relief.

[c] Expected effect is 0.28 × 0.5 = 0.14.

[d] Expected costs are £2000 × 0.28 for patients with pain relief whose pain returns.

From Table 6.1, the ICER is computed for Treatment A versus B as £2381 − £1162/0.544 − 0.418 as £1219/0.126 = £9675 per quality-adjusted life year (QALY).

It is important to note that the word QALY in this calculation does not include the notion of survival or 'years of living', something often associated with a QALY. The treatment benefits are expressed as proportions with or without pain relief. If utilities were available at several time points (e.g. baseline, 3 months and 6 months), the QALY could be a (weighted) average over these time points.

6.3 Markov Modelling/Cohort Simulation

Decision trees are suited to situations in which outcomes occur over short (discrete) periods of time. The decision tree assumes events in each branch are mutually exclusive (i.e. events in one branch cannot occur on the other branch). In some diseases, there are many possible mutually exclusive outcomes, which can make the tree very complicated. For example, in diabetes, treatments can be of several types (insulin, diet, tablets); with each of these treatments, there may be several types of adverse events (AEs); this is followed by complications of illness (such as infections or coma); there may also be another sub-branch for disease progression, resulting in further branches for additional medications. Moreover, if one is interested in repeated analysis at monthly intervals (cycles of time), there is likely to be a 'forest' of branches, and consequently a complex decision tree, which becomes burdensome for later analyses. Therefore, rather than using many possible branches to estimate costs and effects over time, modelling is greatly simplified if we model how patients transition through different health states. A Markov model is particularly well suited to this situation.

6.3.1 Understanding the Transition Matrix

A Markov model in health economic evaluation attempts to model the movement of patients from one health state to another. The objective is to estimate the costs and effects expected to accrue as a result of transitions from varying health states over a period of time. The period of time over which transitions from health states are observed depends on the situation. In situations such as pain relief, this might be a duration (i.e. cycle time) of 1 year. In other diseases, such as cancer, it might be over several years so that expected costs and effects can be determined over the lifetime of patients. Some patients with advanced cancer have short survival times. For example, in lung cancer, in which survival time may be as short as 3 months, the costs could be computed over each week (each cycle is 1 week). In other disease areas with longer survival time, the costs and effects might be calculated over cycles of months or years.

For example, in cancer, patients may move from a health state called 'Progressive Disease' to 'Death' over several months. The probabilities of moving between health states are captured as a matrix of probabilities called a *transition matrix*. It is called a Markov model because the chance of moving to a different health state depends only on the current health state (and not on health states prior to the current state). This is called the Markov property: $X_{t+1} = X_t + \varepsilon$, where X_{t+1} is the health state at time $t + 1$ and is dependent only on the current health state X_t. An example of a 'typical' Markov model for cancer is shown in Figure 6.3.

There are some particularly important concepts in Markov models which need mentioning:

1. *Health states*: These can be simple descriptions, such as 'Death' or 'Disease Progression', or more complicated, such as defining a clinical benefit. For example, for pain relief, a difference of '>2 points at 6 months' may be classified as a health state. However, this definition might need some justification (by a panel of clinical experts).

2. The *transition probability* is the chance of moving between one health state and another. One would need to determine which states a patient can move to. For example, a patient cannot move from a state of 'Death' to 'Progressive Disease', although the reverse is possible.

The source of the transition probability is also important. Where was it taken from? How reliable is it? Is it a conditional probability or a joint probability? For example, rates of AEs are extracted from clinical trial reports. Many AE summary tables report percentages using the denominator of the intent to treat (ITT) or safety population (i.e. the total sample size), which may be joint probabilities, whereas the value that may be of real interest is the probability of the event, given (conditional on) which treatment group patients are in.

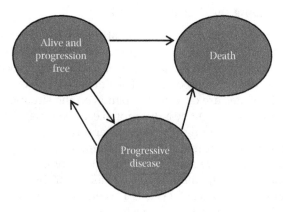

FIGURE 6.3
A simple Markov model applied to cancer.

TABLE 6.2

Example Transition Matrix on Experimental Treatment after 1 Month Post-Randomisation

		Post-Baseline (After Treatment)			
		Mild	Moderate	Death	Total
Baseline	Mild	0.45	0.35	0.20	1
	Moderate	0.05	0.60	0.35	1
	Death	0	0	1	1

An example of a transition matrix for patients with health states Mild, Moderate and Death after 1 month of treatment with an experimental treatment is shown in Table 6.2.

Prior to computation of the transition matrix in Table 6.2, patients would have been randomised to one of two treatments: Experimental or Control. Patients at the start of their treatment are considered to be in either a Mild or a Moderate state (e.g. based on Eastern Cooperative Oncology Group [ECOG] or similar inclusion criteria). The proportion of patients who started the trial at baseline in a Mild health state and then remained in the same state after treatment was 45%. Very few patients (5%) who started the trial in a Moderate state improved into a Mild state post-baseline. In fact, 60% of patients who presented with Moderate disease did not change states after starting treatment. It is assumed that the probability of a patient *moving* from one health state to another is the same (constant) from one future time point to another.

In order to compute what the transition matrix will look like after 1 month, we need to know the initial probabilities (or the probabilities at baseline). These are assumed to be $a_0 = [0.45, 0.35, 0.20]$ (the fact that these are the same as the first row is coincidental).

After 1 month of treatment, therefore, the transition matrix is a_0 multiplied by the 3×3 transition matrix in Table 6.2: $a_1 = [0.45 \times 0.45 + 0.35 \times 0.05 + 0.20 \times 0, 0.45 \times 0.35 + 0.35 \times 0.60 + 0.30 \times 0, 0.45 \times 0.30 + 0.35 \times 0.35 + 0.30 \times 1] = [0.22, 0.37, 0.56]$.

The 1×3 matrix $a_1 = [0.22, 0.37, 0.56]$ describes the probabilities in each of the health states 'Mild', 'Moderate' and 'Death'; 1 month after treatment, this 1×3 matrix now becomes the initial matrix needed for further calculations. After 2 months (cycle 2 of the process), this would be a_1 multiplied by the 3×3 transition matrix, and so on. In general, one can compute the proportion of patients for the n + 1th step in any of these three health states by simply multiplying a_n by the given transition matrix, where a_n is the 1×3 matrix at the current step:

$$a_1 = a_0 * P$$

$$a_2 = a_1 * P$$

$$a_3 = a_2 * P \ldots a_n = a_{n-1} * P$$

where **P** is the transition matrix, and so on. We will see that this is a much simpler way of computing the QALYs in the next example.

6.3.2 Duration of Time

The last important factor associated with the Markov model is the duration of the interval of time, termed a *cycle*. In the example in Section 6.3.1 (Table 6.2), the cycle was 1 month. The cycle could be longer or shorter. For shorter cycles, more computations are required. For example, if we wished to know the probability of health states in patients aged from 18 to 85 years, and we used a cycle duration of 1 month, there would be many computations. It would be easier to define the cycle duration on a yearly basis, so that we end up with 67 transition matrices, with the first (initial) one starting at the age of 18 and the last one at 85 years. If one of the health states were death, we might find at the 67th cycle (85 years) that most patients are in the 'Death' health state.

Example 6.2: Computing Transition Probabilities with SAS

In this example, we will continue with the three health states in Table 6.2 and the same transition matrix as in Section 6.3.1. In addition, we provide the transition matrix for the control group.

First, define the transition matrix and the initial starting matrix. The transition matrix for the experimental arm E is

		Mild	Moderate	Death	Total
	Mild	0.45	0.35	0.20	1
Experimental	**Moderate**	0.05	0.60	0.35	1
	Death	0	0	1	1

The initial matrix $a_{0E} = [0.45, 0.35, 0.20]$.
For the control arm, these are

		Mild	Moderate	Death	Total
	Mild	0.25	0.25	0.50	1
Control	**Moderate**	0.10	0.45	0.45	1
	Death	0	0	1	1

and $a_{0C} = [0.25, 0.25, 0.50]$.

Step 2: Compute the n-step (or n-cycle) transition probabilities. These can be calculated using the SAS code below for the first five cycles. We need to do this for each treatment group separately for 12 cycles.

Cycle 1:

		Mild	Moderate	Death	Total
	Mild	0.45	0.35	0.20	1
Experimental	**Moderate**	0.05	0.60	0.35	1
	Death	0	0	1	1

		Mild	Moderate	Death	Total
	Mild	0.25	0.25	0.50	1
Control	**Moderate**	0.10	0.45	0.45	1
	Death	0	0	1	1

$\mathbf{a}_{0E}$ (Experimental) = [0.45 0.35 0.20]
$\mathbf{a}_{0C}$ (Control) = [0.25 0.15 0.60]

Hence, $\mathbf{a}_{1E} = \mathbf{a}_{0E} \times \mathbf{P}_E$ = [0.45 0.35 0.20] x $\mathbf{P}_E$ = [0.22 0.37 0.41] after cycle 1, where $\mathbf{P}_E$ is the transition matrix for the experimental group. Similarly, for other cycles:

Cycle 2: $\mathbf{a}_{2E}$ = [0.11 0.29 0.59] after cycle 2
Cycle 3: $\mathbf{a}_{3E}$ = [0.07 0.22 0.71] after cycle 3
Cycle 4: $\mathbf{a}_{4E}$ = [0.04 0.15 0.80] after cycle 4
Cycle 5: $\mathbf{a}_{5E}$ = [0.02 0.11 0.87] after cycle 5

```
proc iml;
xinit={0.45 0.35 0.20};
trans={0.45 0.35 0.20,
   0.05 .60 0.35,
      0 0 1};
xaft1=(xinit)*trans; **end of cycle 1**;
xaft2=xaft1*trans; **end of cycle 2**;
xaft3=xaft2*trans; **end of cycle 3**;
xaft4=xaft3*trans; **end of cycle 4**;
xaft5=xaft4*trans; **end of cycle 5**;

print xinit trans xaft1 xaft2 xaft3 xaft4 xaft5;
quit;
```

The SAS code for generating these transition matrices can also be determined using PROC ENTROPY or the code provided here (for the experimental treatment). A simple loop could be added for more cycles.

Note how the proportion of patients increases in the 'Death' state and falls in other states, and also how the probabilities sum to 1. This procedure can be easily repeated for the Control arm, by simply changing the numbers in the code provided here. Once the transition probabilities are estimated, further minor data manipulation in SAS can be carried out to estimate the expected costs and utilities. Costs and utilities can be included in an additional vector and the elements of the matrices multiplied after each cycle to generate the expected costs and QALYs.

Example 6.3: Using Transition Probabilities to Compute QALYs for a Cost-Effectiveness Analysis

Continuing with Example 6.2, it will now be shown how to compute the cost per QALY and the ICER. We will assume that patients were treated for 3 months but followed up for 12 months during the trial. We also require some additional information on costs and utilities for each of the three health states.

First, determine the transition matrix for each treatment group. The transition matrix for the experimental group is

		Mild	Moderate	Death	Total
	Mild	0.45	0.35	0.20	1
Experimental	**Moderate**	0.05	0.60	0.35	1
	Death	0	0	1	1

The utilities and costs for each health state (or overall) need to be stated as

	Utility (EQ-5D)	Monthly Costs (£) Experimental[a]	Monthly Costs (£) Control[a]
Mild	0.78		
Moderate	0.40	954	435
Death	0		

[a] Including drug costs: treatment was for a 3 month period, but follow up was over 12 months.

AE rates for each of the health states can be extracted. It is possible that there are no observed differences in toxicity between treatment groups overall, but differences might exist between groups within health states (e.g. patients with poorer health states on the experimental arm might also have higher toxicities compared with control). Next, determine the number of cycles. In this case, we will use 12 cycles for illustration purposes, where each cycle is 1 month. Hence, the expected costs and QALYs for each month will be generated and then added up.

The proportion of patients in each of the health states will need to be determined *before* the utilities can be applied to generate the QALYs. An initial number is used in the case of a cohort simulation (as in this example). The initial number is arbitrary (such, as 10,000 patients in this cohort). The idea is to estimate the cumulative costs and effects over the 10,000 patients. Hence, total costs and QALYs are given per 10,000. This arbitrary number appears to be a convention (Briggs et al., 2006). The calculations for the experimental arm are shown in Table 6.3 without

TABLE 6.3

Calculation of Markov Trace (Expected Costs and QALYs) for the Experimental Treatment Group

End of Cycle (Month)	Mild	Moderate	Dead	Expected Cost	Proportion Alive	Expected QALY
Initial	10,000				1	
1	2200 (0.22)	3700 (0.37)	4100 (0.41)	$(0.22 + 0.37) \times 954 = 563$	0.59	$(0.22 \times 0.78) + (0.37 \times 0.40) = $ **0.32**
2	1100 (0.11)	2900 (0.29)	5900 (0.59)	$(0.11 + 0.29) \times 954 = 382$	0.41	$(0.11 \times 0.78) + (0.29 \times 0.40) = $ **0.20**
3	700 (0.07)	2200 (0.22)	7100 (0.71)	$(0.07 + 0.22) \times 954 = 277$	0.29	$(0.07 \times 0.78) + (0.22 \times 0.40) = $ **0.14**
4	400 (0.04)	1500 (0.15)	8000 (0.80)	$(0.04 + 0.15) \times 954 = 181$	0.20	$(0.04 \times 0.78) + (0.15 \times 0.40) = $ **0.09**
5	200 (0.02)	1100 (0.11)	8700 (0.87)	$(0.02 + 0.11) \times 954 = 124$	0.13	$(0.02 \times 0.78) + (0.11 \times 0.40) = $ **0.06**
6	200 (0.02)	700 (0.07)	9100 (0.91)	$(0.02 + 0.07) \times 954 = 86$	0.09	$(0.02 \times 0.78) + (0.07 \times 0.40) = $ **0.04**
7	100 (0.01)	500 (0.05)	9400 (0.94)	$(0.01 + 0.05) \times 954 = 57$	0.06	$(0.01 \times 0.78) + (0.05 \times 0.40) = $ **0.028**
8	80 (0.008)	320 (0.03)	9600 (0.96)	$(0.008 + 0.02) \times 954 = 77$	0.04	$(0.008 \times 0.78) + (0.02 \times 0.40) = $ **0.014**
9	52 (0.0052)	233 (0.023)	9715 (0.97)	$(0.0052 + 0.023) \times 954 = 27$	0.03	$(0.0052 \times 0.78) + (0.023 \times 0.40) = $ **0.013**
10	30 (0.003)	160 (0.016)	9810 (0.98)	$(0.003 + 0.016) \times 954 = 18$	0.02	$(0.003 \times 0.78) + (0.016 \times 0.40) = $ **0.0087**
11	20 (0.002)	100 (0.010)	9980 (0.99)	$(0.002 + 0.010) \times 954 = 11$	0.01	$(0.002 \times 0.78) + (0.010 \times 0.40) = $ **0.0056**
12	15 (0.0015)	72 (0.0072)	9913 (0.99)	$(0.0015 + 0.0072) \times 954 = 8$	0.01	$(0.0015 \times 0.78) + (0.007 \times 0.40) = $ **0.0038**
Total				**£1,811**		**0.923 QALYs**

adjustment for discounting. One may discount the expected costs and effects and then sum across the cycles.

In Table 6.3, the total expected costs and QALYs *per patient* are estimated. The costs and QALYs are added up over all 12 time points. Similar calculations are done for the control arm, so we end up with two expected costs and two QALYs (one for each treatment group). If we wanted the total costs for a cohort of 10,000 patients over 12 months, then the values in the expected cost and QALY columns would be multiplied by the actual numbers in each health state. The ICER is derived by calculating the difference in expected costs and dividing by the difference in expected QALYs.

6.3.3 Checking the Model

Checking a Markov model is not like checking a statistical model. In a statistical model, one can check the plots of the residuals against fitted values to check random distribution. In Markov models, however, the assumptions revolve around checking the inputs of the model and asking questions such as

- How reliable are the initial and transition probabilities? Also, do they sum to 1?
- How accurate are the estimated probabilities? Were they based on all the evidence presented?
- How accurate are the estimated unit prices of resources?
- Is the Markov assumption valid: does the probability of moving from the current health state to the next one really depend on the current state (or, in fact, on the entire previous history before the current state)?
- Some tests can be carried out for testing the validity of the Markov assumption by turning a first-order Markov model into a second-order model and observing the impact on the transition probabilities. This test is rarely applied in published health economic models.

Sensitivity analyses (Chapter 7) are often carried out that ask several 'What if'–type questions, for example 'What if the probabilities were higher/lower – how would it impact the ICER?'

6.4 Analysis of Patient-Level Data

In Example 6.3, data from a clinical trial were summarised, and a Markov model was then used to estimate the ICER for the study treatments delivered over 3 months, but patients were followed up over 12 months. Hence, costs and effects were assessed over 12 months. Example 6.3 used summaries of data rather than patient-level data. We now look at an example using patient-level data, called a *patient-level cost-effectiveness analysis* or a *stochastic analysis*. In this section, we show how to summarise the results from a lung cancer trial.

6.4.1 Patient-Level Costs

It is important prior to analysis to determine what we wish to assume about the distribution of costs and effects. In Chapter 5, costs were shown to have a very skewed distribution. We can model patient-level costs by first computing the total cost per patient for each group. Subsequently, the patient-level costs can be modelled with a gamma model, and one can estimate the mean costs in each treatment group.

6.4.2 Patient-Level Effects (Utilities and Efficacy)

Unlike costs, which are computed for each patient over the time horizon of interest, utility data are collected at several time points for analysis. Efficacy data are also collected at several time points. Therefore, we need a way to model repeated data with a view to deriving a single value, which is the mean QALY. For utility data, several options are possible, such as repeated measures analysis, computing area under the curve (AUC) for each patient or computing the (model-based) mean estimates of utility at each time point. There is no need to compute differences in mean utility, because the differences in mean QALY are more relevant. In any case, mean differences in utility between treatments are not easy to interpret even for the most experienced clinician or health economist. Once mean utilities are estimated, these are used to adjust the efficacy outcomes. If the efficacy outcomes are continuous and available at each time point, the utilities can be used to weight the efficacy outcomes. We now show an example from a published lung cancer trial.

Example 6.4: Case Study of a Patient-Level Cost-Effectiveness Analysis

Background

This example is taken from Khan et al. (2015) and Lee (2012) using data from a trial in 670 lung cancer patients randomised to either Erlotinib (n = 350) or Placebo (n = 320). One interesting feature of this analysis was that a pre-specified subgroup of patients who developed skin rash after receiving the first cycle of treatment had a particularly marked treatment effect (Lee, 2012). Therefore, a cost-effectiveness analysis was also carried out for the subgroup. Full details of this cost-effectiveness analysis are given in Khan et al. (2015), and it is briefly described here.

The primary end point was overall survival (OS). Progression-free survival (PFS) was a secondary end point. EQ-5D utilities were collected from baseline at monthly intervals during the first year. Treatment benefit (efficacy) was not demonstrated either statistically or clinically for the primary end point. However, for the subgroup of patients with rash, there was evidence of some benefit. In practice, one would not bother to do a cost-effectiveness analysis in all the patients. However, it is useful to do this in both the whole group and the subgroup to show the impact and the change in conclusions.

Patients were followed up until progression or death. The primary end point was OS; secondary end points were PFS, safety and HRQoL. Pre-specified subgroup analyses included whether or not patients developed treatment-related rash within the first 28 days (first-cycle rash). HRQoL using EQ-5D was assessed monthly. Randomised patients who took study medication were included in this analysis. Rash was assessed using a subjective five-point scale: no rash (grade 0), erythema alone (grade A), erythema with papules (grade B), erythema with papules and pustules (grade C), and erythema with papules and confluent pustules (grade D).

6.4.2.1 Treatments

The main comparison is patients randomised to Erlotinib who developed first-cycle rash (ER) versus SC (Placebo group) receiving palliative radiotherapy (RT) in UK participating sites. There was no recommended standard treatment for this patient population; hence, SC was considered the most suitable comparator. Patients received Erlotinib (150 mg) or matching Placebo daily until progression. Dose reductions to 100 or 50 mg were allowed. Patients who took Erlotinib but did not develop rash (ENR) are discussed in the sensitivity analyses.

6.4.2.2 Costs/Resources

The costs relevant for this analysis were costs from the drug (Erlotinib), RT, additional anti-cancer treatments, patient management (hospital clinic visits, day cases and admissions) and managing important treatment-related AEs (e.g. diarrhoea and rash). Resource use was collected monthly on the case report forms (CRFs). Unit prices were taken from hospital pharmacy records, published National Institute for Health and Care Excellence (NICE) reports (where available), published literature, National Health Service (NHS) reference costs and the British National Formulary (BNF, 2012). Costs were estimated in pounds sterling. No discounting was applied to costs or health benefits <1 year; discounting at 3.5% per annum was applied in the second year; nearly all patients (>99%) had progressed or died by 2 years.

6.4.2.3 Drug Costs

Erlotinib price (taken from the BNF) was set at £54.37/tablet for 150 mg, £44.12 for 100 mg and £25.21 for 50 mg. Drug use was determined from the recorded number of tablets dispensed and returned. Drug cost per patient was estimating by multiplying the duration of Erlotinib use by the unit price. For Placebo, drug cost was set to 0. Where additional treatments were given (for palliation), unit costs were identified.

6.4.2.4 Supportive Care

The mean price of palliative RT for advanced stage non-small cell lung cancer patients (including RT planning) was assumed to be £120 per visit using NHS reference prices (year 2011–2012). The total RT cost is computed by multiplying the duration of RT by the unit price. Costs for additional anti-cancer treatments were included in the analysis: the daily price of gemcitabine/carboplatin was assumed to be £47.50, and for vinorelbine it was £19.18.

6.4.2.5 Clinic Visits/Admissions

Resource use (hospital and additional clinic visits, day cases, night stays) were recorded on the CRFs. Only hospitalisations from AEs that were recorded as definitely/probably/possibly treatment related were used to compute costs of hospital overnight stays. The price/day was £100 for a clinic visit, £670 for a day case and an additional £730 for overnight stay (i.e. £1400 including admission).

6.4.2.6 AEs

Clinically important grade 3 and above serious adverse events (SAEs) with >5% frequency were rash, diarrhoea and dyspnoea; other toxicity rates (<1%) were similar between treatments, and expected cost differences were negligible. Patients with the highest AE grade (maximum AE grade) were included; it was assumed that palliative treatment was taken, even if AEs were milder. For rash and diarrhoea, the costs per day were set to £4.30 and £8.59, respectively; the cost of morphine at 15 mL/day is about £0.22; steroid use (dexamethasone) was based on £13.80 per 100 tablets with standard doses of three tablets of 4–8 mg per day at £0.42/daily dose; the cost of salbutamol was assumed to be £0.13/daily dose.

Duration of AEs was computed from their start/end dates; daily unit prices of medications for treating AEs (following UK practice) were computed based on a monthly course. The total cost (per patient) was computed by adding component costs: drug cost, supportive care (RT), additional clinic visits, hospital day cases, hospital admissions and treatments for SAEs. Total costs were determined and modelled using a generalised linear model assuming a gamma distribution to derive the incremental mean costs. If the total cost was 0, a small increment of 0.001 was added for modelling purposes (because costs need to be strictly >0 for a gamma model to be applied).

6.4.2.7 Utilities

The EQ-5D was used to construct health utilities for economic evaluation. The EQ-5D consists of five scales (mobility, self-care, usual activities, pain/discomfort and anxiety/depression) collected from baseline until progression/death. Responses were converted into utilities using a UK social tariff based on the time trade-off (TTO) method. Missing utility data were handled through multiple imputation (MI) techniques. Responses to EQ-5D were captured on paper CRFs during clinic assessments (monthly in year 1 and 6 monthly thereafter).

The EQ-5D in this trial was collected at multiple time points (at baseline and monthly thereafter for the first year, or until death or progression). For each patient, it may also be possible to analyse the data based on one of the following approaches:

(i)   Compute the mean utility for each patient across all post-baseline time points, taking into account utilities prior to progression and post-progression. The individual (averaged) patient-level utilities could then be used to adjust the patient-level PFS and post-progression survival (PPS) to get patient-level QALYs.

(ii) Model the data and compute predicted per patient utilities. Here, a repeated measures model could be used, adjusting for baseline to estimate at the patient level the predicted utilities for time points prior to progression and also post-progression. Again, the patient-level predicted utilities for pre- and post-progression can be used to adjust the patient-level PFS and PPS.

(iii) Use a repeated measured linear mixed effects model to adjust for baseline and whether the observed utility occurred pre- or post-progression.

Mean utilities were computed across patients using a baseline adjusted model (such as an analysis of covariance with terms for pre- and post-progression). This method is slightly different in that we use summarised (between patient) utilities from a model. We then also compute mean PFS and PPS and estimate the QALY at the aggregate level. Whatever the choice from (i) to (iii), the objective is to derive a mean utility score for each treatment group to adjust the survival times.

6.4.2.8 Statistical and Cost-Effectiveness Analysis

A patient-level cost–utility analysis was undertaken. The OS, PFS and PPS were determined for each patient. Mean EQ-5D utilities over time were

TABLE 6.4

Data Structure for Patient-Level Cost-Effectiveness Analysis

Patient ID	Treatment	Assessment Month	EQ-5D Utility	OS (Days)	PFS	PPS	Cost
1	Erlotinib	Baseline	0.3	27	20	7	4000
1	Erlotinib	Month 1	0.4	27	20	7	4000
1	Erlotinib	Month 2	0.5	27	20	7	4000
etc.	etc.	etc.	etc.	etc.	etc.	etc.	etc.
2	Placebo	Baseline	0.6	35	10	15	6000
2	Placebo	Month 1	0.2	35	10	15	6000
2	Placebo	Month 2	0.3	35	10	15	6000
etc.	etc.	etc.	etc.	etc.	etc.	etc.	etc.

estimated for pre- and post-progression periods and multiplied by corresponding survival times to derive QALYs for each patient. The total costs were then modelled to derive mean costs. Finally, mean incremental costs, QALYs and the ICER were derived.

An example of the data structure for two patients is shown in Table 6.4 for a patient-level analysis. The mean effects will comprise OS, which consists of PFS and PPS. For example, if a patient lived for 6 months, with a PFS of 4 months, the PPS would be 6 − 4 = 2 months (Table 6.4). This can be calculated if all patients have died; however, when they have not (censored), the OS will be unknown for these patients. Therefore, an alternative approach would be needed, which would model survival data (see Chapter 10). In this trial, all patients had sadly died, and therefore the computations are simplified. For each patient, the PFS and PPS are multiplied by the pre- and post-utility values.

Mean OS and PFS are determined using Kaplan–Meier methods. The mean OS, PFS and PPS are first estimated and are subsequently adjusted by the mean utilities. If patients are still alive at the end of the study, the approach to computing the mean is considerably more complicated. In this case, a form of extrapolation is required to estimate survival rates until a specified age (e.g. 100 years, since few or no patients are expected to be alive at that time). The estimated mean area under the extrapolated curve is determined (for each treatment). The utilities are then applied to the mean survival times to estimate the QALY. In this example, however, no extrapolation is required, Chapter 10 offers SAS code for methods of extrapolation, including how to use more novel methods such as cubic splines and flexible parametric models.

6.4.3 Results

From a total of 670 patients (350 Erlotinib; 320 Placebo), a pre-specified subgroup of 302 Erlotinib versus 278 Placebo were evaluable for first-cycle rash (i.e. after 1 cycle of treatment, patients were monitored if they developed any rash – a side effect of treatment. Since some patients died within the first cycle of treatment – about 28 days – these were considered unevaluable for rash). The main comparison of interest in this cost-effectiveness analysis is the subgroup Erlotinib rash (ER) versus Placebo, because overall, there were no differences between Erlotinib and Placebo: 178/302 (59%) developed rash in the first cycle (ER group), 124/302 (41%) took Erlotinib and did not develop rash in the first cycle (ENR group) and 5/313 (2%) on Placebo had rash.

6.4.3.1 Costs/Resources

Erlotinib was taken as 150 mg tablets by about 83% of patients without any dose reductions; 15% and 2% of patients reduced dose to 100 and 50 mg, respectively. In the rash subgroup, this percentage was 79% for 150 mg tablets, 20% for 100 mg and 1% for 50 mg tablets, respectively. Hence, after taking into account dose reductions and dose delays, the mean cost of Erlotinib was £6863 overall and £7544 in the rash subgroup (no drug costs for Placebo). Table 6.5 shows a summary of the costs and effects in the rash subgroup. All estimates were determined from patient-level data (i.e. for each patient a resource use/no resource use outcome was determined and multiplied by the unit cost).

6.4.3.2 Efficacy

The means and standard errors (SE) were 7.08 (0.48) versus 6.41 (0.44) months for OS and 4.95 (0.36) versus 3.80 (0.29) months for PFS for Erlotinib and Placebo, respectively. In the rash subgroup, OS was 9.08 (0.65) versus 6.91 (0.43) months (Table 6.5); PFS was 6.22 (0.51) versus 4.19 (0.32) months for Erlotinib and Placebo, respectively.

6.4.3.3 Utilities and QALYs

The mean QALY was 0.365 versus 0.3303 overall, yielding an incremental QALY of 0.035. In the rash subgroup, the mean QALY was 0.467 versus 0.337 for Erlotinib versus Placebo, yielding a statistically significant mean incremental QALY of 0.139 (95% confidence interval [CI]: 0.0341, 0.2359; p-value = .0070), in favour of Erlotinib. The improved QALY within the rash subgroup appears to be due to improved survival (Table 6.6), notably PFS. Hence, the mean incremental cost was £7090 and the overall ICER was £202,571/QALY. In the rash subgroup, the mean incremental cost was £7,891 (95% CI £6,999–£8,783), but with better HRQoL/utility, resulting in a base case ICER of £56,770/QALY (Table 6.5). The incremental cost, excluding Erlotinib cost, was £347, and the ICER was £2496/QALY in the rash subgroup.

The mean incremental net benefit (INB) for Erlotinib is not realised until one is prepared to pay in excess of about £202,571 overall and £56,770 in the rash subgroup.

TABLE 6.5

Summary of Cost-Utility Analysis Using Patient-Level Data

Costs Item	Estimated Unit Price (£)	Rash Subgroup		
		Erlotinib (Rash) (£) N = 178 Mean (SE)	Placebo/SC (£) N = 278 Mean (SE)	Difference (p-value) (£)
Erlotinib[a]	54.37/tablet	7544 (764)	0	7,544
Supportive care:				
Palliative RT[b]	120/visit	302 (52)	235 (27)	67 (p = .0449)
Additional treatment[c]	See note c	182 (65)	270 (99)	−88 (p = .85)
Patient Management				
Hospital clinic visit[d]	100/visit	629 (57)	624 (53)	5 (p = 46)
Hospital day case[d]	670/day case	274 (65)	323 (131)	−49 (p = .69)
Hospital admission[d]	730/night	744 (163)	475 (134)	269 (p = .0352)
Adverse events[e]	See note e	221 (34)	114 (27)	107 (p < .001)
Total mean cost (SE)[f] (95% CI)		9,949 (724) (8,530–11,368)	2058 (185) (1695–2420)	7891 p < .001
Incremental cost (SE)[f] (95% CI)		7,891 (614) (95% CI: 6,999–8,783)		

[a] Cost as £1631.53 for 30 tablets (150 mg tablet) or depending on dose; +Unit price based on national NHS tariff (NICE report 2011).

[b] Palliative RT: Diagnosis and treatment of lung cancer update (2011).

[c] On the Placebo arm, seven patients took additional chemotherapy (carboplatin/gemcitabine [n = 5], erlotinib [n = 2]) after progression; on the Erlotinib arm, patients took carboplatin (n = 2), vinorelbine (n = 2) or fragmin (n = 1).

[d] Additional clinic visits and day visits irrespective of reason; unit prices taken from NHS reference costs 2010–2011; nights spent in hospital as a result of treatment-related SAEs.

[e] Total costs for diarrhoea, rash and dyspnoea: duration of each AE was computed from date of onset of the event to date resolved; morphine dose of 15 mL/daily is about £0.22/day; steroid use (dexamethasone) based on £13.84 per 100 tablets and taking three tablets of 4–8 mg per day gives £0.42/daily dose; inhaler: salbutamol, £0.13/daily dose: mean diarrhoea costs were £14 versus £2; mean rash costs were £68 versus £14.90 and mean dyspnoea costs were £139 versus £97.

[f] Determined using a generalised linear mixed model assuming gamma distributed costs for Erlotinib + Rash versus Placebo.

TABLE 6.6

Model Inputs: Effectiveness Measures

	Overall	
	Erlotinib (N = 334)	Placebo (N = 313)
Mean OS (months [SE])	7.08 (0.48)	6.41 (0.44)
Mean PFS (months [SE])	4.95 (0.36)	3.80 (0.29)
Mean PPS (months [SE])	2.13 (0.250)	2.61 (0.236)
Hazard ratio (OS) (95% CI; p-value)	0.92 (0.79–1.08; p-value = .32)	
Hazard ratio (PFS) (95% CI; p-value)	0.81 (0.70–0.95; p-value = .0102)	
Utilities		
Pre-progression EQ-5D (mean [SE])	0.6482 (0.009)	0.6438 (0.011)
Post-progression EQ-5D (mean [SE])	0.5517 (0.016)	0.5760 (0.014)
QALY (years)[a]	0.365 (0.0272)	0.3303 (0.0245)
Incremental QALY (mean [SE])[b]	0.035 (0.0163)	

[a] This is computed as (pre-progression utility) × PFS + (post-progression utility) × PPS.
[b] Erlotinib versus Placebo.

6.5 Patient-Level Simulation

In this section we describe how, given estimates of parameters, such as hazard ratios, mean differences and variances of effects, one can simulate the costs and effects from particular distributions to perform an economic evaluation. There are three main types of simulation in economic evaluation:

(i) Simulating data from aggregate values
(ii) Simulating data from a patient-level data set (e.g. bootstrapping)
(iii) Simulating correlated patient-level data from a patient-level data set

In (i), parameter estimates such as mean differences and standard deviations are used to simulate data. For example, if the mean EQ-5D utility is 0.3 with a standard deviation (SD) of 0.14, 1000 values might be simulated such that the mean and SD of the 1000 values approximate 0.3 and 0.14, respectively.

For (ii), simulation revolves around simulated patient profiles. For example, in a data set of 400 patients with baseline and post-baseline utility data, a random sample of 400 are selected with replacement. This involves repeated sampling with replacement of a particular size (e.g. a sample size of 100 from a total of 400, repeated many times). The correlations between repeated measures and multiple variables are often considered constant during resampling. The third (iii) form of simulation is a more complicated form of simulation, which follows from (ii) in that the 400 patients with repeated data are simulated such that all variables (utilities, costs) take into account correlations between them.

A further type of simulation might involve simulating specific inputs of, say, a Markov model and then using each simulated input into the model to estimate the ICER. For example, the transition matrix probabilities can be simulated such that there might be 1000 possible transition matrices. For each transition matrix, the Markov model is run to estimate the ICER. Therefore, 1000 simulated ICERs might be possible. This latter form of simulation may take a long time, which is often a criticism of a patient-level (or micro) simulation model.

A patient-level simulation can be confused with a probabilistic sensitivity analysis (PSA). The distinction is that the objective of a PSA is to quantify the uncertainty of the observed ICER in relation to a threshold cost-effectiveness ratio. The mechanism by which this uncertainty can be assessed is simulation. A patient-level simulation model, however, is a means of expressing how (single) patients might pass through a certain healthcare pathway. Since patients are considered to be relatively homogeneous (as in clinical trials), each simulated patient profile with specific characteristics (e.g. age 65 years, weight 56 kg, responder to treatment) expresses possible realisations from a specific population. For example, one simulated set of data might result in a patient with an age of 61 years, weight of 56 kg, PFS of 2.1 months and a utility value of 0.14. The outcomes from the (simulated) realisations are subjected to a cost-effectiveness analysis. Table 6.7 summarises the types of simulation approaches for economic evaluation.

Example 6.5: SAS Code for Simulating Multivariate Normal Data (Correlated Patient-Level Data)

Assuming a data set with eight variables, four of which are costs (X_1, X_2, X_3 and X_4) and four effects (Y_1, Y_2, Y_3 and Y_4), the following SAS code uses the observed means of each of the eight variables ($Y_1...Y_4$, $X_1...X_4$) and their covariance (correlations) matrix to simulate 10,000 observations from a multivariate normal distribution.

```
**simulate from group 1 separately**;
**data set holds 4 costs and 4 effects variables**;
data x1;
set coseff;
if group=1;
run;
```

TABLE 6.7

Summary of Simulation Methods

Source for Simulation	Example of Data Used for Simulation	Simulation Result/Outcome
Aggregate	Single mean = 2, SD = 0.9	A list of numbers with a mean = 2 and a SD = 0.9
Patient-level	Baseline, post-baseline measures	A data set with correlated costs and effects – using multivariate simulation
Patient-level	Complete patient profiles	Simulate patients with replacement – such as bootstrapping

```
**generate covariance matrix**;

proc corr data=x1 cov out=cov1;
var x1 x2 x3 x4 y1 y2 y3 y4; *use the observed correlation
structure**;
run;                          *to simulate data**;

data cov11;
set cov1;
if _type_ in ('COV');
drop _type_ _name_;
run;

proc iml;
mean={46.77, 60.13, 67.27, 71.98,
78.07, 84.81, 89.09,91.43}; **mean costs and effects for
group 1**;
use cov11;
read all;
cov1=x1||x2|| x3|| x4|| y1|| y2|| y3|| y4;
call randseed(5760);
do n=1 to 10000; **10,000 simulations**;
x=randnormal(n, mean, cov1);
end;
*smean=mean(x);
*scov=cov(x);
y=x;
varNames = "x1":"x8";
create boot1 from y [colname=varNames];
append from y;
close boot1;
```

6.6 Other Issues in Modelling

A useful set of guidelines are available for decision analytic modelling
(Phillips et al., 2004 and also the NICE 2013 guideline), which underline
important considerations before undertaking economic evaluation. In
particular, a justification for the model choice (decision tree, Markov or
patient-level) should be clearly made. Where patient-level data are available, a decision tree or Markov model might be harder to justify. However,
if it is necessary to model the time course of the disease and expected
costs and effects over a lifetime, modelling becomes essential, especially
when the follow-up in the clinical trial is limited. In this case, a form of
extrapolation or some way of estimating the proportion of patients in the
distant future will need to be estimated along with the expected costs
and effects.

If any structural assumptions are made in the models, these should also
be stated. For example, any assumptions about costs need to be justified.

TABLE 6.8

Model Inputs: Modelling Situations

Model	Data Type/Source
Decision tree	Aggregate/RCTs/published
Markov/cohort simulation	Aggregate or summarised from studies: RCTs/published
Patient-level/stochastic	Patient-level costs and effects collected: RCTs
Discrete event simulation	Patient-level data or aggregate data RCTs
Evidence synthesis	Patient-level data and aggregate data RCTs/real-world evidence

The costs of AEs, for example, might only include those AEs that have a high probability of relationship to treatment. It is possible that there are many unrelated AEs on a comparator arm, which might result in a much higher/ lower incremental cost and a higher/lower ICER. A further issue is correcting for the beginning and end of the cycle. It might be assumed when using a Markov model that transitions between health states occur at the beginning or end of a cycle only, which may not be the case (transitions can occur at any time).

One further model will be discussed in Chapter 9: the evidence synthesis model, in which evidence from several sources (direct and indirect) is used together to build a network of information to estimate treatment effects. These are often Bayesian models for indirect and mixed treatment comparisons. A summary of the type of modelling in economic evaluation is shown in Table 6.8.

Exercises for Chapter 6

1. Describe the differences between a decision tree, a Markov model and a stochastic analysis when performing an economic evaluation for a clinical trial.

2. What is the difference between a conditional and a joint probability from the perspective of undertaking a Markov model for an economic evaluation?

3. What is the difference between bootstrapping, a patient-level simulation model and simulating correlated patient-level data?

4. What is a patient-level simulation? Describe a patient-level simulation for a cancer trial in which the survival times are assumed to be exponentially distributed.

5. Explain the process involved when simulating data from data that are multivariate normally distributed.

Appendix 6A: SAS/STATA Code

6A.1 Simulating Correlated Multivariate Normal Data

6A.1.1 SAS

Assuming a data set with eight variables, four of which are costs (X_1, X_2, X_3 and X_4) and four effects (Y_1, Y_2, Y_3 and Y_4), the following code uses the means and covariance matrix to simulate 10,000 observations from a multivariate normal distribution.

```
**simulate from group 1 separately**;
**data set coseff holds 4 costs and 4 effects variables**;
data x1;
set coseff;
if group=1;
run;


**generate covariance matrix**;

proc corr data=x1 cov out=cov1;
var x1 x2 x3 x4 y1 y2 y3 y4;
run;

data cov11;
set cov1;
if _type_ in ('COV');
drop _type_ _name_;
run;

proc iml;
mean={46.77, 60.13, 67.27, 71.98,
78.07, 84.81, 89.09,91.43}; **mean costs and effects for group
1**;
use cov11;
read all;
cov1=x1||x2|| x3|| x4|| y1|| y2|| y3|| y4;
call randseed(5760);
do n=1 to 10000; **10,000 simulations**;
x=randnormal(n, mean, cov1);
end;
*smean=mean(x);
*scov=cov(x);
y=x;
varNames = "x1":"x8";
create boot1 from y [colname=varNames];
append from y;
close boot1;
```

6A.2 STATA Code

We wish to draw a sample of 1000 observations from a normal distribution N(M;V), where M is the mean matrix of three variables: EQ-5D, Effects and Costs – all normally distributed. Assume also that **V** is the covariance matrix. Simulation would occur for each treatment group separately.

```
. matrix M = 4, -6, 0.7
. matrix V = (8, 5, 2 \ 7, 3, 1 \ 1, 1, 1)
. matrix list M
M[1,3]
c1 c2 c3
r1 4 -6 .7
. matrix list V
symmetric V[3,3]
c1 c2 c3
r1 8
r2 7 3
r3 1 1 1
. drawnorm EQ5D Effects Costs, n(1000) cov(V) means(M)
```

7

Sensitivity Analyses

7.1 Introduction to Sensitivity Analysis

The purpose of sensitivity analysis in health economic evaluation is to assess the uncertainty of conclusions (such as those based on the incremental cost-effectiveness ratio [ICER]) by varying the model inputs. Data from a single clinical trial might be of limited value, especially if it is the only trial that has been carried out for the disease under investigation (or if it is unlikely that a similar trial will be conducted again in the foreseeable future). For example, assuming that the treatment-related adverse event rate is 5% for a particular treatment group, but the reliability of the rate is questionable (because, for example, the trial was unblinded, and hence any assessment of treatment relatedness is biased), one might assess the impact on the ICER if the adverse event rate were larger or smaller than 5%.

In Markov and decision-tree-type models, uncertainty is assessed by techniques such as *one-way sensitivity analyses* or *two-way sensitivity analyses*. In one-way sensitivity analysis, inputs (costs or effects) are varied by a certain amount one at a time. For example, the costs might be varied by ±10% and the impact on the ICER observed while all other inputs are held constant. Then, separately, another factor is varied, and again the impact on the ICER is observed. In two-way sensitivity analysis, inputs are simultaneous. For example, the adverse event rate might increase by 10% and utilities be reduced by 10% (at the same time). The effect on the ICER (i.e. how it has changed from the base case value) is then observed. One may also vary several inputs, but this can become more complicated. An alternative approach might be to work out the simultaneous percentage changes in two or more inputs for a required ICER (i.e. working backwards to a desired ICER).

In sensitivity analysis, the decision on how much to vary the inputs is arbitrary. One could use 95% (or even 80%) confidence intervals (CIs) to describe the uncertainty of input variables. For example, one could calculate the 95% CI for the mean utility for a given treatment group and also report the 95% CI for the mean costs. The upper and lower limits could then be used to assess the impact on the ICER. The problem here is that the upper and lower 95% CIs of inputs may result in extreme ICERs. An alternative approach might be to simply present the 95% CI for the mean ICER using Fieller's theorem. However, the plausible range of values for the confidence interval can lead to different interpretations, especially if the confidence interval covers two regions in the cost-effectiveness plane. For example, if the 95% CI for the ICER ranges from −£456 to +£899 for a mean ICER of £678/quality-adjusted life year (QALY), there are two possible conclusions, which might influence health policy when comparing a new treatment with a standard. On the one hand, the new treatment is more expensive but more effective (£899); on the other hand, the new treatment is less effective but cheaper (−£456). Therefore, the ICER needs to be interpreted with caution, because we need to know the position of incremental costs and effects (numerator and denominator) in the cost-effectiveness plane. In this example, if the cost-effectiveness threshold was £1000, both decisions could be implemented. An alternative approach is to avoid CIs and simply estimate the chance that the new treatment is cost-effective, as in probabilistic sensitivity analysis (PSA). In this chapter, examples of one- and two-way sensitivity analysis will be shown, followed by examples of frequentist and Bayesian PSA.

7.2 One-Way Sensitivity Analysis

Example 7.1: One-Way Sensitivity Analysis

Assume the mean costs for Treatments A and B are £3000 and £4000, respectively, with mean utilities of 1.1 and 1.2. This gives a base case ICER of (£4,000 − £3,000)/0.1 = £10,000/QALY for B versus A. The base case ICER is the reported ICER prior to any sensitivity analysis. The impact on the ICER from one-way sensitivity analyses is shown in Table 7.1. Note, in the first row, how all other factors (costs for Treatment B,

TABLE 7.1

One-Way Sensitivity Analysis for Example 7.1

	ICER
Base Case	**£10,000**
Costs Treatment A + 10%	£7,000[a]
Costs Treatment B − 10%	£6,000[b]
Utility Treatment A + 10%	£9,090[c]

[a] A 10% increase of £3000 = £3300, so that the revised ICER = £4000 − £3300/0.1 = £7000.
[b] A 10% decrease of £4000 = £3600, so that the revised ICER = £3600 − £3000/0.1 = £6000.
[c] A 10% increase in utility of 1.1 = 1.21, so that the revised ICER = £4000 − £3000/0.11 = £9090.

utilities for A and B) remain the same, while only costs for Treatment A are increased by 10%. Reducing the cost of Treatment B by 10% has the greatest impact on the ICER.

7.3 Two-Way Sensitivity Analysis

Example 7.2: Two-Way Sensitivity Analysis

Following on from Example 7.1, we now show the impact on the ICER when two or more factors are varied simultaneously. In the first few rows of Table 7.2, we note that costs and utilities have been varied by a 10% increase for Treatment A, while simultaneously we assume that costs and utilities for Treatment B have decreased by 10%, resulting in a revised ICER of £2307/QALY. In other words, if the costs of Treatment A were to rise by 10% and quality of life also improved by a similar amount, but Treatment B costs were to fall along with QALYs by 10%, Treatment B would be cheaper but also less effective. In the second row, the reverse is assumed: Treatment A costs and utilities are assumed to fall by 10%, whereas for Treatment B, they are assumed to increase. In this case, the ICER rises to over £5000/QALY.

In Table 7.1, the ICER fell by 30%–40% as a result of varying drug costs by ±10%. One way of displaying the impact on the ICER is through the use of a type of plot called a *tornado diagram*. These are essentially horizontal bar charts that attempt to show the influence of the factors that impact the ICER.

An example of a tornado or influence plot is shown in Figure 7.1 for some simulated data for two treatments A and B (not the same as in Example 7.2). The SAS code to generate the tornado diagram is shown in Appendix 7A.1. It is apparent from the plot that varying Treatment A costs by ±10% does not

TABLE 7.2

Two-Way Sensitivity for Example 7.2

Treatment	Factor	Change (%)	ICER (Base Case £10,000)
A	Costs	+10	−£2,307
A	Utility	+10	
B	Costs	−10	
B	Utility	−10	
A	Costs	−10	£5,151
A	Utility	−10	
B	Costs	+10	
B	Utility	+10	

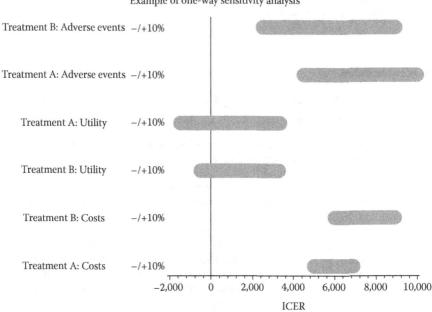

FIGURE 7.1
Tornado plot for one-way sensitivity analysis.

impact the ICER, whereas varying the costs of treating adverse events on Treatment B by the same amount results in large changes in the ICER.

7.4 PSA

It is sometimes (wrongly) assumed that PSA is a Bayesian method for expressing uncertainty about the chance of cost-effectiveness of a new treatment. The frequentist approach uses Monte-Carlo simulation (Bayesian methods require a prior distribution and use Markov chain Monte-Carlo methods to approximate complex integrations) to estimate the (long-run) probability of cost-effectiveness for varying cost-effectiveness thresholds. The approach is as follows:

1. Identify the inputs to vary (e.g. cost, utilities, adverse events).
2. Identify parameters associated with the inputs (e.g. mean, standard deviation, rates).
3. Identify the correlation or covariance matrix of the variables from which to simulate, if required.

4. Determine whether the Monte-Carlo simulation will be from a univariate or a multivariate distribution.
5. Determine whether a multivariate simulation will take into account the 'mixed' nature of the distributions (i.e. will the simulation assume data are multivariate normal or a combination of normally and non-normally distributed data?).
6. Write SAS/STATS code to simulate.
7. Compute the ICER for each simulation.
8. Determine the proportion of ICERs above a certain cost-effectiveness threshold value (λ).
9. Plot the cost-effectiveness acceptability curve (CEAC).
10. Do the above for each treatment group separately (or all together if technically possible).

Example 7.3: PSA

This example is taken from the earlier example of Lee (2012) for a lung cancer trial. The following costs and effects were simulated from a multivariate distribution using the SAS IML procedure.

1. Total costs: assumed to be normally distributed
2. Utilities: assumed to be beta binomial (pre- and post-progression separately)
3. Overall and progression-free survival: assumed to be exponentially distributed

Hence, four components of the ICER were to be simulated. The objective was to simulate from a multivariate distribution in which variables (inputs) had different distributions. The mean and standard deviation for each cost component are shown in Table 7.3 for the erlotinib treatment group.

Note that it would be incorrect to simulate these independently and then merge them together. What is needed is a correlation or covariance matrix.

TABLE 7.3

Distributional Assumptions of Inputs for Simulation

	Distribution Assumed
Total costs	Gamma
EQ-5D pre-progression utility	Beta binomial
EQ-5D post-progression utility	Beta binomial
OS	Exponential
PFS	Exponential

Note: OS: overall survival; PFS: progression-free survival.

TABLE 7.4

Correlation Matrix Used for Simulating Multivariate Data
(Experimental Arm Only)

	Total Costs	PrP Utility	PP Utility	OS	PFS
Total costs	1				
Pre-utility	0.61	1			
Post-utility	0.65	0.89	1		
OS	0.71	0.58	0.62	1	
PFS	0.69	0.55	0.66	0.88	1

Note: OS: overall survival; PFS: progression-free survival; PrP: pre-progression; PP: post-progression.

We assume that the five components have the correlation matrix shown in Table 7.4.

Simulating multivariate data from mixed distributions can be very complex. Several methods exist, and we will describe one of them here, which uses the Fleishman power transformation method (1978). A method using copulas is also possible, and interested researchers may wish to explore this. Simulation using copulas was successfully accomplished in Khan et al. (2015).

The process of simulating multivariate correlated data from any distribution requires the use of Fleishman coefficients for pairwise variables to generate the *intermediate* correlations. For example, with two inputs, total costs and pre-progression utility, these can be relabelled as X_1 and X_2, respectively. Hence

$$X_1 = a_1 + b_1 Z_1 + c_1 Z^2_1 + d_1 Z^3_1 \tag{7.1}$$

$$X_2 = a_2 + b_2 Z_2 + c_2 Z^2_2 + d_2 Z^3_2 \tag{7.2}$$

The values of a_i, b_i, c_i and d_i for ($i = 1, 2$) are estimated through the power transform, and may be found in Fleishman tables using higher-order moments. Once the values of a_i, b_i, c_i and d_i are estimated, the intermediate correlation between X_1 and X_2 is determined by using the equation from Vale & Maurelli (1983):

$$R_{x_1 x_2} = \rho(b_1 b_2 + 3b_1 d_2 + 3d_1 b_2 + 9d_1 d_2) + \rho^2(2c_1 c_2) + \rho^3(6d_1 d_2)$$

This is then repeated for each of the pairwise variables to form an 'intermediate correlation matrix'. SAS code based on Fan et al. (2002) can be used to simulate multivariate correlated data where each variable is from *any* distribution (Appendix 7A.2).

The SAS code in 7A.2 uses PROC IML for a Newton Raphson iterative procedure to find the values of a, b, c and d in Equations 7.1 and 7.2. Using the data from Lee et al. (2012), with a sample size of 670, a data set of n = 670 patients was simulated with corresponding costs and effects. This was then repeated 10,000 times (10,000 data sets of size n = 670 with n = 350 on

TABLE 7.5

Output from Multivariate Simulation Using the Fleishman Method for the Experimental Treatment Group

Patient	Total Costs (£)	Pre-Utility	Post-Utility	PFS (months)	OS (months)
1	4000	0.4	0.2	1.2	1.8
2	8000	0.8	0.4	2.2	4.4
3	3500	0.8	0.4	1.1	2.7
etc.	..	....	...	...	...
350	8500	0.7	0.2	4.9	6.2

Note: OS: overall survival; PFS: progression-free survival.

TABLE 7.6

Simulated ICERs from Each Data Set

Simulation	Mean Cost (Erlotinib)	Mean Cost (Placebo)	Mean QALY (Erlotinib)	Mean QALY (Placebo)	ICER
1	6,700	5,200	1.80	1.77	50,000
2	8,300	5,200	1.60	1.59	310,000
3	11,700	10,100	1.10	1.0	16,000
etc.					
10,000					

erlotinib and n = 320 on placebo). For each data set, the ICER was computed. An example extract of three simulated data points for three patients is shown for the experimental treatment group in Table 7.5.

For each data set, the mean, costs and QALYs are calculated. It is important that the way the mean costs are computed is the same as the base case approach. For example, if the mean costs were estimated using a generalised gamma model, this should be repeated for each data set rather than using the simple mean (particularly if covariates are to be included in the simulation exercise). The simulated ICERs for each data set will look something like those in Table 7.6.

The next step will be to determine the proportion of ICERs below a specified cost-effectiveness threshold (or willingness to pay). Recall from Chapter 2 that the desirable ICER is

$$Dc / De < \Delta$$

where λ is the cost-effectiveness threshold.

In 10,000 simulated ICERs, the proportion that is below λ is recorded and then plotted to generate the CEAC. An alternative expression is

$$\lambda \times \Delta_e - \Delta_c$$

the INMB introduced in Section 2.2.

TABLE 7.7

Proportion of ICERs below the CE Threshold for First Three Simulated ICERs

Simulation	ICER (£)	CE Threshold (£)	Is ICER >CE Threshold?	Probability ICER <CE Threshold
1	50,000	1,000	No	0
2	310,000	1,000	No	0
3	16,000	1,000	No	0
1	50,000	2,000	No	0
2	310,000	2,000	No	0
3	16,000	2,000	No	0
		etc.		
1	50,000	20,000	No	0
2	310,000	20,000	No	0
3	16,000	20,000	Yes	1/3

Note: CE: cost-effectiveness.

What we require is the probability of INB to be >0 for specified values of λ.

In Table 7.7, the proportion of times that the ICER is less than the cost-effectiveness threshold of £1000 and £2000 is zero. As the cost-effectiveness increases to £20,000, one out of the three simulated ICERs is below £20,000. Hence, the probability of cost-effectiveness at a threshold of £20,000 is 33% (in this example). The process continues, and then the probabilities are plotted on the y-axis and the cost-effectiveness threshold on the x-axis. The CEAC for the erlotinib cost-effectiveness trial (Khan et al. 2015) is shown in Figure 7.2.

An example of SAS code used to generate a CEAC assuming data are multivariate normal is shown in Appendix 7A.3.

7.5 Bayesian Sensitivity Analyses

In Examples 7.2 and 7.3, Monte-Carlo simulation was used to estimate the (long-run) probability of cost-effectiveness for varying levels of the cost-effectiveness threshold. The Bayesian approach is likely to lead to the same conclusions, depending on prior beliefs about parameters and distributions of cost and effects. For example, if the prior distributions are non-informative (i.e. observed data dominate the results over the prior beliefs), the frequentist and Bayesian analysis will give very similar results. Different choices for priors can lead to different cost-effectiveness conclusions (Maiwenn et al., 2000).

Whether the analysis is frequentist or Bayesian, the objective is to quantify uncertainty around the ICER. In Section 2.2, we saw the use of Fieller's

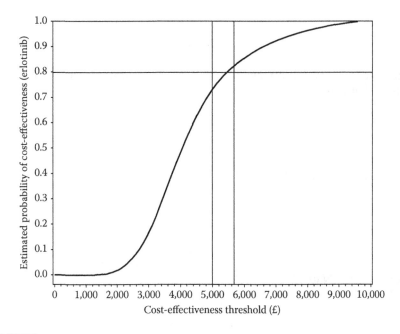

FIGURE 7.2
CEAC showing probability of cost-effectiveness of erlotinib versus placebo for the data in Table 7.7.

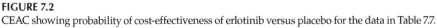

theorem and CIs for expressing this uncertainty. Fieller's method and boot-strapping do not use prior information, and statistical inferences are based only on the observed data. No assumptions are made on the distribution of costs and effects.

The interpretation of the CI from a Bayesian analysis is called a *credible interval* (CrI). The CrI expresses uncertainty about the (posterior) point estimates ICER and is interpreted as offering a probability of where the parameter of interest lies: 'there is a 95% probability that the interval will contain the true ICER'. Compare this with the more wordy frequentist 95% CI: 'if you repeated this analysis 1000 times, then 95% of the confidence intervals would contain the true ICER'.

PSA under a Bayesian context is best understood starting with the bivariate general linear model (GLM). We start with each patient having two responses: one for costs, c_i, and one for effects, e_i, contained in the matrix **Y**. In addition, there are two treatment groups (although it can be extended to more than two groups). This is the matrix **X**, in which the first column is the intercept and the second column an indicator variable for the treatment group. We also have the parameter matrix β and the error vector ε.

Hence, $Y_{ij} = \mu + \tau_i + \varepsilon_{ij}$ is a standard form of a GLM, where the subscripts $i = 1, 2$ for each treatment group and j are the observations for each patient in the trial. Furthermore, μ is a vector

$$\begin{pmatrix} \mu_c \\ \mu_e \end{pmatrix}$$

of the mean costs and effects, τ is a vector of treatment effects for each of costs and effects

$$\begin{pmatrix} \tau_c \\ \tau_e \end{pmatrix}$$

and ε_{ij} is a matrix of residual errors:

$$\begin{pmatrix} \varepsilon_{11} & \varepsilon_{21} \\ \vdots & \vdots \\ \varepsilon_{1n} & \varepsilon_{2n} \end{pmatrix}$$

where $\varepsilon_{ij} \sim \text{MVN}(0, \Sigma)$
   If **Y** is

$$\begin{pmatrix} c_{11} & e_{21} \\ \vdots & \vdots \\ c_{1n} & e_{2n} \end{pmatrix}$$

a matrix (of $n \times 2$) of responses for costs and effects, then the model (in a frequentist framework) can be written as

Y =		X		β +		ε	
Costs	**Effects**	**Intercept**	**Group**	**Parameters**		**Error**	
2000	1.2	1	1	$\mu_1 \, \mu_2$		ε_{11}	ε_{21}
3000	1.6	1	1	$\tau_{11} \, \tau_{12}$		ε_{12}	ε_{22}
8000	0.8	1	1			ε_{13}	ε_{23}
etc.	etc.	.	.			.	.
		.	.			.	.
		.	0			.	.
		1	0			ε_{1n}	ε_{2n}

The parameters to be estimated are

$$\mu = \begin{pmatrix} \mu_c \\ \mu_e \end{pmatrix}$$

and the variance–covariance matrix Σ.

In a Bayesian context, we assume prior distributions for

$$\mu = \begin{pmatrix} \mu_c \\ \mu_e \end{pmatrix}$$

which we can call

$$\mu_0 = \begin{pmatrix} \mu_{0c} \\ \mu_{0e} \end{pmatrix}$$

and also a prior distribution for Σ, termed

$$\Sigma_0 = \begin{pmatrix} \sigma_c^2 & \rho\sigma_c\sigma_e \\ \rho\sigma_c\sigma_e & \sigma_e^2 \end{pmatrix}$$

The objective is to combine the prior information with the likelihood functions to determine an (updated) estimate of the mean costs and effects. These are called *posterior means*. Using the posterior means, the (posterior) ICER is then derived. Simulations from the posterior distributions (i.e. posterior mean costs and effects) are carried out (similarly to the frequentist method), resulting in 10,000 (for example) posterior mean costs and effects. With these data, the CEAC can be generated as before.

Bayesian modelling is often carried out in WINBUGS software; however, the PROC MCMC and a PROC GENMOD in SAS are also available for Bayesian analysis.

Example 7.4: Bayesian Sensitivity Analysis

Using the (artificial) data set with a sample size of n = 10 (five in each group) for illustration purposes only, we can perform a Bayesian PSA in SAS (Table 7.8).

First, a standard GLM bivariate model (also called a multivariate analysis of variance [MANOVA] model) is coded in SAS as follows:

```
PROC GLM data=example;
Class treatment;
Model costs effects = treatment;
Means treatment;
Run;
```

This code generates the mean costs for Treatments A and B as £3600 and £2000, respectively. The mean effects are 1.0 and 0.8 QALYs, respectively, and hence the ICER is £3600 − £2000/(1.0 − 0.8) = 1400/0.2 = £7000/QALY.

TABLE 7.8

Example Data Set Used in Bayesian Analysis

Patient	Costs	Effects	Treatment
1	2000	1.2	A
2	3000	1.6	A
3	8000	0.8	A
4	3000	0.8	A
5	2000	0.6	A
6	3000	0.8	B
7	1000	0.9	B
8	1500	0.7	B
9	2400	0.8	B
10	2100	0.8	B

In a Bayesian analysis, the data can be analysed as a joint bivariate function or as two separate models for costs and effects. The disadvantage of using two separate models (one for costs and one for effects) is that it is necessary to ignore the correlation between costs and effects. For illustrative purposes, we can use two separate Bayesian regression models. The SAS code can be easily modified for a joint model taking correlation into account:

$$\text{Costs} = \beta_{0c} + \beta_{1c} * \text{Treatment} \quad \text{Effects} = \beta_{0e} + \beta_{1e} * \text{Treatment}$$

We will assume that β_{0c}, β_{1c}, β_{0e} and β_{1e} have normal prior distributions such that

$\pi(\beta_1) \propto 1$ (dropping the subscript as we assume the same priors)

$\pi(\beta_0) \propto 1$ for costs and effects, and further

$\pi(\sigma) \sim \text{gamma}(\text{shape}= 0.001; \text{scale} = 0.001)$

These are also the default priors in PROC GENMOD (called flat or non-informative priors). The SAS code for a Bayesian analysis is found in Appendices 7A.2 and 7A.3.

The 10,000 posterior estimates (in the SAS code) are sent to the data sets 'postc' and 'poste', which can be merged and the posterior ICERs derived. The mean costs for Treatments A and B and the mean ICER are very similar to those of the MANOVA analysis. The CEAC based on posterior means can be determined in a similar way to that described in Example 7.3. The interested reader can compare the results from this Bayesian analysis with simulations from a multivariate mixed distribution (using, for example, the Fleishman power transformation) as an exercise.

A 95% CrI for the ICER can also be generated once we have the posterior means, using the fact that differences of two independent normal random variables are also normally distributed (or using the CEAC).

7.6 Issues in Interpreting and Reporting Results from Sensitivity Analysis

Table 7.9 provides a summary of how uncertainty analysis is presented, with some key issues.

7.6.1 Interpreting the CEAC and Cost-Effectiveness Plane

Figure 7.3 shows the CEAC and scatter plot on the cost-effectiveness plane from the same trial as discussed in Example 6.4 (Khan et al., 2015; Lee et al., 2012). The CEAC shows the long-term 'probability' of cost-effectiveness for varying cost-effectiveness thresholds. The meaning of 'cost-effectiveness' here is the proportion of times we can expect the INB to be >0. If one reads across from the y-axis at 5% and 95%, the corresponding cost-effectiveness thresholds are about £26,000 and £72,000. That is, there is a 95% probability that the true ICER will lie between £26,000 and £72,000. In Figure 7.3, at a cost-effectiveness threshold of £30,000/QALY (read along the x-axis), the corresponding value on the y-axis is about 15%. Hence, we can expect about 15% of ICERs to lie below £30,000. At some very large-value cost-effectiveness thresholds there will be a 100% chance that the new treatment is

TABLE 7.9

Summary of Main Issues in Uncertainty Analyses

Method	Situation/Model	Main Issues
One-way	Decision model	Arbitrary amounts varied Simple Can result in extreme ICERs
Two-way	Decision model	Arbitrary amounts varied Simple but can become complicated with >2 inputs
Confidence or credible intervals/ bootstrap CIs	Patient-level data	Based on statistical theory Can result in extreme ICERs Can lead to different conclusions In some cases may not be estimable
CEAC	Any	Simple to interpret Provides an intuitive approach to estimating the probability of cost-effectiveness A bit more complicated with >2 treatments Unclear how secondary end points information is incorporated
Scatter plot/CE plane	Any	Shows uncertainty around the individual ICERs Can see visually how many ICERs might lie in different quadrants
CE frontier	Any	More appropriate when comparing >2 treatments

Note: CE: cost-effectiveness.

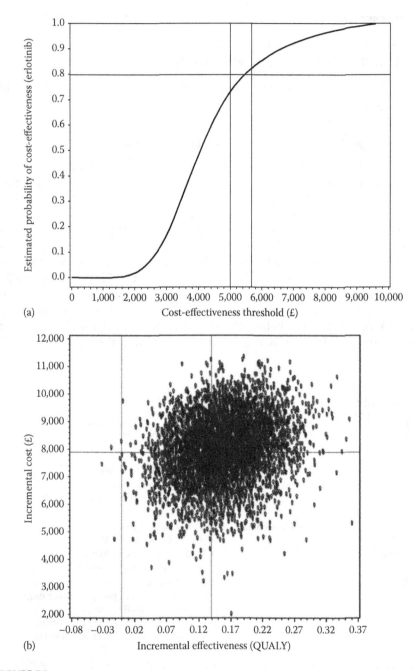

FIGURE 7.3
(a) CEAC and (b) cost-effectiveness plane for cost-effectiveness of erlotinib. (From Khan et al. *BMJ.*, 5, E006733, 2015. With Permission.)

cost-effective, because ultimately, if there is a very large budget (money is not a concern), any new treatment is likely to be cost-effective.

There is no universal agreement as to what probability level is acceptable for a new treatment to be cost-effective. The general rule is the higher the probability, the better. An 80% probability of the new treatment being cost-effective at a threshold of £30,000 has been reported. However, in practice, other factors, such as disease and number of alternative treatments available, might also play a part, and not just a single estimate of the probability of cost-effectiveness. A separate question about the cost-effectiveness threshold is also worth mentioning. The UK criterion (National Institute for Health and Care Excellence [NICE] threshold) is £20,000 to £30,000, but recently NICE has increased the threshold from £30,000 to £50,000 under specific conditions (e.g. end of life treatments).

The CEAC threshold also is often based on the primary end point of a trial. Separate CEACs for subgroups and secondary end points could be derived. How these are interpreted all together is not immediately clear. For example, consider a CEAC for the primary end point of a new treatment which is modestly cost-effective (e.g. 75% chance of cost-effectiveness at a threshold of £30,000/QALY). If all secondary end points generated very strong effects, the CEAC for the primary end point would not change. That is, a decision based on the CEAC might be the same whether a handful of secondary efficacy end points showed strong improvements in effects or whether they showed worsening with the new treatment. For this reason, decisions should not be made on the CEAC alone unless the CEAC in some way uses secondary efficacy end points (not to mention the correlation between primary and secondary efficacy end points).

The scatter plot in Figure 7.3b shows the distribution of ICERs. There are reference lines at zero and observed incremental costs and effects. Since most of the values (ICERs) are to the right of the reference line of zero, on average, the new treatment shows promising signs of effectiveness in this subgroup of patients (the subgroup who developed rash in the first cycle). However, the incremental costs exhibit considerable uncertainty (values lie between £2,000 and £11,000). These costs may not appear numerically very large, but since the ICER is a ratio, when the incremental costs are divided by the incremental QALYs, the ICER is large.

In a Bayesian analysis, the CEAC would be almost identical to Figure 7.3. However, the y-axis of Figure 7.3 would represent a true probability of cost-effectiveness and not a relative frequency probability, although for practical purposes, the two methods reach the same conclusions. The CrI, however, is a Bayesian concept, which answers a slightly different question than the frequentist CI. If a PSA is conducted under a Bayesian framework (e.g. using a non-informative prior), the CEAC generated can be used to estimate the 95% CrIs. The 95% CrIs suggest that there is a 95% probability that the ICER will lie between £26,000 and £45,000. The probability of the new treatment being cost-effective at a threshold of £30,000 is 58%. The upper limit of this 95% CrI includes the value of £30,000; therefore, there is a still a 95% chance that the ICER might be cost-effective at a cost-effectiveness threshold of £30,000.

Exercises for Chapter 7

1. What is the difference between uncertainty and variability?
2. Explain the differences between one-way, two-way and PSA in the context of economic evaluation.
3. 'We don't need uncertainty analysis, we can just use confidence intervals'. Criticise this statement.
4. For the data in Table 7.1, use the SAS code provided to construct a tornado diagram after adjusting costs by 5% and utilities by 15%.
5. What is the CEAC? Explain how the algorithm is generated when the correlation between costs and effects is negligible. It is decided that costs and effects are likely to be correlated, which should be reflected in the PSA. What would you do differently now when generating the CEAC?
6. For the data in Table 7.8, generate the CEAC.

Appendix 7A: SAS/STATA Code

7A.1 SAS Code for Generating a Tornado Plot for a One-Way Sensitivity Analysis

```
%let refvalue=0;
%let x_min=-2000;
%let x_max=15000;

data tornado_data;
input text_label $ 1-35 min max;
length=max-min;
datalines;
Treatment A: Costs  -/+10% 5000 6900
Treatment A: Utility  -/+10% -1500 3300
Treatment A: Adverse Events -/+10% 4500 9945
Treatment B: Costs  -/+10% 6000 8900
Treatment B: Utility  -/+10% -500 3300
```

```
Treatment B: Adverse Events -/+10% 2500 8945
;
run;

/* determine the order, based on length of lines */
proc sort data=tornado_data out=tornado_data;
by length;
run;
data tornado_data; set tornado_data;
stack_order=_n_;
run;

proc sql;
create table foo as
select unique stack_order as start, text_label as label
from tornado_data;
quit; run;
data control; set foo;
fmtname = 'torfmt';
type = 'N';
end = START;
run;
proc format lib=work cntlin=control;
run;

/* Insert missing value between each line segment, to use with
'skipmiss' */
data tornado_data; set tornado_data;
y=stack_order; ICER=min; output;
y=stack_order; ICER=max; output;
y=.; ICER=.; output;
run;

goptions device=png;
goptions noborder;

ODS LISTING CLOSE;
ODS HTML path=odsout body="&name..htm"
  (title=" ") style=htmlblue;

goptions gunit=pct ftitle="albany amt/bold" htitle=3
ftext="albany amt" htext=3;

symbol1 interpol=join color=cx00ff00 value=none width=25;

axis1 label=none major=none minor=none style=0 offset=(3,3);
axis2 c=black label=none minor=none order=(&x_min to &x_max
by 1000 h=1.5);
```

```
title1 ls=1.5 "Tornado Plot";
title2 h=2.5 "Example 7.1: Tornado Plot - One Way Sensitivity
Analysis";

proc gplot data=tornado_data;
format stack_order torfmt.;
plot stack_order*ICER /
 skipmiss
 vaxis=axis1
 haxis=axis2
 noframe
 href=&refvalue
 des='' name="&name";
run;

quit;
ODS HTML CLOSE;
ODS LISTING;
```

7A.2 SAS Code for Fleishman Coefficients: Simulating Multivariate Data from Any Distribution

Before one computes the Fleishman coefficients for multivariate simulation, measures of skewness and kurtosis are needed. Once these are obtained (e.g. using PROC UNIVARIATE), the following IML code is used to determine the Fleishman coefficients. The following code adapted from Fan et al. (2002) is provided to generate Fleishman coefficients for three variables: cost, progression-free survival (PFS) and utility.

```
PROC IML;
SKURT ={ -1 3.5, 3 8, 0 1, -2 6};

START NEWTON ;
RUN FUN;
DO ITER = 1 TO MAXITER
WHILE (MAX(ABS(F))> CONVERGE) ;
RUN DERIV; DELTA=-SOLVE(J,F);
COEFF=COEFF+DELTA; RUN FUN;END;
FINISH NEWTON;
MAXITER=50;CONVERGE=0.0001;

START FUN;
COST=COEFF[1];**x1;
UTILITY=COEFF[2];**x2;
PFS=COEFF[3]; **x3**;

F=(COST**2+6*COST*PFS+2*UTILITY**2+15*PFS**2-1)//
  (2*UTILITY*(COST**2+24*COST*PFS+105*PFS**2+2)-SKEWNESS)//
  (24*(COST*PFS*UTILITY*(1+COST**2+28*COST*PFS)+PFS**2+2)*
```

```
(12+48*COST*PFS+141*UTILITY**2+225*UTILITY**2))-KURTOSIS;
FINISH RUN;

START DERIV;
J=((2*COST+6*PFS)||(4*UTILITY)||(6*COST+30*PFS))//
 ((4*UTILITY*(COST+12*PFS))||(2*COST**2+24*COST*PFS+105*
PFS**2+2))
 ||(4*UTILITY*(12*COST+105*PFS)))//
((24*(PFS+UTILITY**2*(2*COST+28*PFS)+48*PFS**3))||
(48*UTILITY*(1+COST**2+28*COST*PFS)+141*PFS**2))||
(24*(COST+28*COST*UTILITY**2+2*PFS*(12+48*COST*PFS+141*UTILITY
**2+225*PFS**2)+PFS**2*(48*COST+450*PFS)))));
FINISH DERIV;
DO; NUM =NROW(SKURT);
DO VAR= 1 TO NUM;
SKEWNESS=SKURT[VAR,1];
KURTOSIS=SKURT[VAR,2];
COEFF={1,0, 0.0, 0.0};
RUN NEWTON;
COEFF=COEFF`;
SK_KUR=SKURT[VAR,];
COMBINE=SK_KUR||COEFF;
IF VAR=1 THEN RESULT=COMBINE;
ELSE IF VAR >1 THEN RESULT=RESULT//COMBINE;
END;
PRINT 'Y = A+BX +CX^2 +DX^3'; ** prints the coefficients of

          Y = a+bZ+cZ^2+dZ^3 power transformation**;
```

Once the coefficients have been determined, the IML SAS code similar to that in Appendix 6A.1 can be used to simulate the multivariate distributions of PFS, utility and costs:

```
ods graphics on;
proc genmod data=ce;
class treat;
model cost = treat/ dist=normal cl;
bayes seed=45 outpost=postc diagnostics=all summary=all;
run;
ods graphics off;

data postc1(keep=iteration costa intcost);
set postc;
rename treata=costa intercept=intcost;
run;

ods graphics on;
proc genmod data=ce;
class treat;
model eff = treat/ dist=normal cl;
```

```
bayes seed=87 outpost=poste diagnostics=all summary=all;
run;
ods graphics off;

data poste1(keep=iteration effa inteff);
set poste;
rename treata=effa intercept=inteff;
run;
data all;
merge postc1 poste1;
by iteration;
icer=costa/effa;
run;
```

7A.3 SAS Code/Algorithm for Generating CEAC and CE for Data in Khan et al. (2015)

```
**for Group 1; erlotinib+Rash group generate the correlation
between the six variables**;

**get correlations for total costs pre and post progression
survival times OS, overall survival, progression free survival
and post-progression survival**;

proc corr data=costs2 out=corr1 ;
where grash1_num=3; **Erlotinib group**;
var total_costs mean_eqos_pre mean_eqos_post os pfs pps;
run;

proc means data=costs2 n mean std skew kurt;
var total_costs mean_eqos_pre mean_eqos_post os pfs pps;
run;

data senserlot;
set corr1 ;
if _type_='CORR';
run;

**start process for simulating multivariate data**;

proc factor data=senserlot n=6 outstat=facout;
run;

data pattern;
set facout;
if _type_='PATTERN';
drop _type_ _name_;
run;

proc means data=costs2;
class grash1_txt;
```

```
var total_costs mean_eqos_pre mean_eqos_post os pfs pps;
run;

**USE IML to simulate multivariate data**;

proc iml ;
use pattern ;
read all var _num_ into F;
F=F`;

data=rannor(J(10000,5,0)); **10,000 simulations**;
data=data`;
Z=F*data;
Z=Z`;
x1=Z[,1]*9736+10290; ** linear transformation for a specified
mean and SD**;
x2=Z[,2]*SD1+mean1; **specify values for the mean and SD
here**;

x3=Z[,3]*SD2+mean2;
x4=Z[,4]*SD3+mean3;
x5=Z[,5]*SD4+mean4;
x6=Z[,6]*SD53+mean5;

Z=x1||x2||x3||x4||x5||x6;

create A from Z [COLNAME={x1 x2 x3 x4 x5 x6}];
append from Z;
quit;

proc means data=a n mean std skew kurt;
var x1 x2 x3 x4 x5 x6;
run;

proc means data=costs2 n mean std skew kurt;
class grash1_txt ;
var total_costs mean_eqos_pre mean_eqos_post os pfs pps;
run;

**data manipulation**;

data simer1;
set a;
rename x1=total_costs ;
rename x2=mean_eqos_pre;
rename x3=mean_eqos_post;
rename x4=os;
rename x5=pfs;
rename x6=pps;
sim=_n_;
sim2=sim;
```

```
diff=sim-sim2;
run;

data simerl2;
set simerl;
qaly1=mean_eqos_pre*pfs;
qaly2=mean_eqos_post*os;
totalqaly=(qaly1+qaly2)/12;
run;

**get mean costs and effects**;

proc univariate data=simerl2 noprint;
by diff2;
output out=simcost mean=meancostTR;
var total_costs;
run;

proc univariate data=simerl2 noprint;
by diff2;
output out=simqaly mean=meanqalyTR;
var totalqaly;
run;

data sim3TR;
merge simcost simqaly;
by diff2;
run;

***now repeat for the other treatment group: Placebo+Rash**;
proc corr data=costs2 out=corr2 ;
where grash1_num=1;
var total_costs mean_eqos_pre mean_eqos_post os pfs pps;
run;

**for placebo +rash**;

data sensplac;
set corr1 ;
if _type_='CORR';
run;

proc factor data=sensplac n=6 outstat=facoutpl;
run;

data patternpl;
set facoutpl;
if _type_='PATTERN';
drop _type_ _name_;
run;
```

```
proc means data=costs2;
class grash1_txt;
var total_costs mean_eqos_pre mean_eqos_post os pfs pps;
run;

proc iml ;
use patternpl ;
read all var _num_ into F;
F=F`;

data=rannor(J(10000,5,0));
data=data`;
Z=F*data;
Z=Z`;
x1=Z[,1]*2818+5468;
x2=Z[,2]*0.283+0.524;
x3=Z[,3]*0.240+0.574;
x4=Z[,4]*7.254+6.916;
x5=Z[,5]*5.279+4.192;
x6=Z[,6]*4.434+2.723;

Z=x1||x2||x3||x4||x5||x6;

create placr from Z [COLNAME={x1 x2 x3 x4 x5 x6}];
append from Z;
quit;

proc means data=placr n mean std skew kurt;
var x1 x2 x3 x4 x5 x6;
run;

proc means data=costs2 n mean std skew kurt;
class grash1_txt ;
var total_costs mean_eqos_pre mean_eqos_post os pfs pps;
run;

data simplr;
set placr;
rename x1=total_costs ;
rename x2=mean_eqos_pre;
rename x3=mean_eqos_post;
rename x4=os;
rename x5=pfs;
rename x6=pps;
sim=_n_;
sim2=mod(sim,279);
diff=sim-sim2;
diff2=diff/279;
run;
```

```
data simplr2;
set simplr;
if diff2=1000 then delete;
qaly1=mean_eqos_pre*pfs;
qaly2=mean_eqos_post*os;
totalqaly=(qaly1+qaly2)/12;
run;

proc univariate data=simplr2 noprint;
by diff2;
output out=simcostplr mean=meancostPR;
var total_costs;
run;

proc univariate data=simplr2 noprint;
by diff2;
output out=simqalyplr mean=meanqalyPR;
var totalqaly;
run;

data sim3plR;
merge simcostplr simqalyplr;
by diff2;
run;

data psa1;
merge sim3tr sim3plr;
by diff2;
incost=(meancostTR-meancostPR);
ineff=meanqalyTR-meanqalyPR;
icer=incost/ineff;
rename diff2=sim;
run;

data icer2;
set psa1;
do thresh=1000 to 100000 by 1000;
output;
end;
run;

proc sort data=icer2;
by thresh sim;
run;

data icer3;
set icer2;
if icer <= thresh then count=1;
else count=0;
run;
```

```
proc means data=icer3 n mean stderr lclm uclm;
var icer;
run;

proc freq data=icer3 noprint ;
table count*thresh / out=icer3;
run;

data icer4;
set icer3;
if count=1;
run;

**generating the CEAC **;

 data anno;
 function='move';
 xsys='1'; ysys='1';
 x=0; y=0;
 output;

 function='draw';
 xsys='1'; ysys='1';
 color='vibg';
 x=100; y=100;
 output;
run;

**plot CEAC**;

goptions device=cgmof971 gsfmode=replace gsfname=gsasfile
reset=global
 ftext = swiss htitle = 5 htext = 3 gunit = pct
border cback = white hsize = 10in vsize = 6in;

axis1 order = (0 to 1 by 0.1) offset = (3,0) label = (a=90
'Probability of Cost Effectiveness (Erlotinib)') width=2
minor=none
axis2 order = (0 to 100000 by 1000) offset = (5,5) label=('CE
Threshold (£)') width=2 minor=1000 major=10000;**y axis**;

proc gplot data=icer4;
symbol1 interpol=line value=none width=2.5 height=1.3
color=black repeat=6;
plot percent*thresh / c=black frame vaxis = axis1 haxis=axis2
href=30000;

label thresh='CE Threshold (£)';
run;
```

```
**plot CE plane **;

goptions anno=ano1 device=cgmof971 gsfmode=replace
gsfname=gsasfile reset=global
 ftext = swiss htitle = 5 htext = 3 gunit = pct
border cback = white hsize = 10in vsize = 6in;

axis1 order = (10000 to 60000 by 10000) offset = (3,0) label
= (a=90 'incremental cost') width=2 minor=none; **y aaxis**;
axis2 order = (-0.1 to 2 by 0.1 ) offset = (5,5) label=('inc
eff') width=2 minor=1000 major=10000;
proc gplot data=icer4;
plot incost*ineff='*' / href=0 vref=0;
run;
```

8

Sample Size and Value of Information for Cost-Effectiveness Trials*

8.1 Introduction

This chapter contains some sections of a more technical nature. Sections 8.1 through 8.3 are more practical, with some sample size formulae and several examples. Sections 8.4 and 8.5 focus on the factors affecting the sample size for cost-effectiveness trials presented in the context of both frequentist and Bayesian methodology. Section 8.6 provides examples of sample sizes in a Bayesian context, and Sections 8.7 and 8.8 discuss further advanced issues. However, despite complex Bayesian computations, a Windows-based application (SCET-VA-version 1.0®) is available free of charge from the author on request. For researchers who wish to simply calculate the sample size required for specific inputs, without wanting to know more about mathematical derivations, Sections 8.1 through 8.3 are adequate. Section 8.9 covers value of information (VOI) methods.

8.2 Sample Sizes for Cost-Effectiveness

Traditionally, sample size calculations are carried out to design clinical trials to detect clinically relevant differences between two treatments. In clinical trials designed for economic evaluation, it is usual to collect patient-level data on resource use (costs) in addition to efficacy and safety. While there may be arguments for powering a clinical trial for demonstrating clinical benefit, the arguments for powering for cost-effectiveness have not been fully appreciated. Several reasons why calculating a sample size for cost-effectiveness may not be considered relevant are discussed.

---

* Contributions towards Sections 8.1 through 8.9 of this chapter were made by Dr. Shah Jalal Sarker (BARTS Cancer Institute, Queen Mary College) and Professor Anne Whitehead (Lancaster University)

First, it is argued that without showing a clinical effect, there is unlikely to be any cost-effectiveness argument; therefore, efforts should be directed towards the clinical end point. Second, although the magnitude of a clinical effect can be justified, the size of a measure of 'effectiveness' may be less clear. For example, in trials where the end point is a combination of quality of life and quantity of life, the composite 'economically relevant' difference in quality-adjusted life years (QALY) is usually unknown. Third, ethics committees rarely query a protocol with a cost-effectiveness objective but without a sample size justification, whereas they would almost certainly query a protocol without a sample size justification for the primary clinical outcome. Fourth, it may not be considered ethical to power a trial purely for demonstrating cost-effectiveness, because it is construed as an economic argument and not a clinical argument for the trial.

Despite these reasons, the risks of a new experimental treatment which 'misses' the opportunity to demonstrate value either for reimbursement purposes or for national health policy reasons are real and should not be underestimated. This is similar to a 'false negative' in that the likelihood of showing cost-effectiveness is real, but the trial may not have been designed appropriately to demonstrate this. It is perhaps for this reason that funding bodies such as the Cancer Research UK Clinical Trials Awards and Advisory Committee (CTAAC), the National Institute for Health Research (NIHR) and Health Technology Assessment (HTA) awards bodies have sections on sample sizes for cost-effectiveness objectives in their funding applications. Moreover, some regulatory agencies, such as in Australia and Canada, require cost-effectiveness data before granting reimbursement of a new health technology/drug. In the United Kingdom, the National Institute for Health and Care Excellence (NICE) offers an advisory service which provides advice to optimise designs of clinical trials aimed at demonstrating cost-effectiveness. The ability to power for cost-effectiveness, therefore, allows researchers, policymakers, funders and drug companies to plan for another level of uncertainty on the path of licensing and reimbursement or changing the standard of care in an economically viable way.

8.3 Sample Size Methods for Efficacy

In the past, sample sizes for cost-effectiveness were calculated in different ways. One method was to calculate sample sizes for clinical effects and costs separately, and then choose the greater of the two. However, it is unclear what constitutes an important difference in costs between treatment groups. Moreover, this approach treated costs and effects as independent. In practice, patients with smaller treatment effects could have larger costs because they might be treated with other, more expensive treatments, or they might have stopped taking the treatment (e.g. due to lack of effect or to side effects), which might also result in increased costs.

In clinical trials of efficacy, the sample size depends on several variables, including the hypothesised clinically relevant difference (e.g. measured using a hazard ratio) and the variability associated with the clinical effect. A sample size formula might look like Equation 8.1, used for continuous measures (as in a two-sample t-test):

$$n_E = (z_\alpha + z_\beta)^2 \frac{(\sigma_{E1}^2 + \sigma_{E2}^2)}{\Delta_{ER}^2} \tag{8.1}$$

where:
  n   is the sample size for each treatment group
  σ_{E1}^2 and σ_{E2}^2 are expressions of the variability of the response in each of the two groups
  σ_{ER}^2 is the clinically relevant difference
  z_α and z_β are special values taken from published tables (or software) which express the risk of a false positive (often 5%) or a false negative (often 20%)

For example, for a drug which is expected to improve the Hba1c value by 0.9 mm/L with a variance of 3.1 in group 1 and 3.6 in group 2, the sample size required would be

$$n = (1.96 + 1.28)2 \times (3.1 + 3.6)/0.92$$

$$= 10.496 \times 6.7/0.81$$

$$= 87 \text{ per group (or 174 patients in total)}$$

8.4 Sample Size Formulae for Cost-Effectiveness: Examples

In Equation 8.1 and calculations, there is no mention of costs for any of the treatment groups or a (co)-relationship between costs and clinical effects, nor are there details of a cost-effectiveness threshold, or anything about a cost-effectiveness ratio or benefit. When cost-effectiveness is also considered in the same trial (a cost-effectiveness trial), the sample size formula needs to include additional factors:

(a) The way the effectiveness element is evaluated – that is, using the incremental cost-effectiveness ratio (ICER) approach or the incremental net benefit (INB) approach
(b) Costs in each group
(c) Variability of costs in each group
(d) Correlation between costs and effects in each group
(e) Cost-effectiveness threshold

Taking these factors into account makes the sample size formula more complicated. In some situations (such as when using Bayesian methods), a formula, even a complex formula, may not be available. The sample size in these situations is calculated using simulation methods. Although the sample size problem can be approached through simulations (Al et al., 1998), it is better (easier) to determine sample sizes by using a sample size formula rather than having to do lots of programming. In a clinical trial for comparing two groups, in which costs and effects are collected prospectively, we propose the following sample size formula (Sarkar & Whitehead, 2004) when patient-level data are available:

$$n_1 = \left(z_\alpha + z_\beta\right)^2 \frac{\left[r\left(k^2\sigma_{E1}^2 + \sigma_{C1}^2 - 2k\rho_1\sigma_{E1}\sigma_{C1}\right) + \left(k^2\sigma_{E2}^2 + \sigma_{C2}^2 - 2k\rho_2\sigma_{E2}\sigma_{C2}\right)\right]}{r\left[k\Delta_{ER} - \Delta_{CR}\right]^2}$$

$$n_2 = \left(z_\alpha + z_\beta\right)^2 \frac{\left[r\left(k^2\sigma_{E1}^2 + \sigma_{C1}^2 - 2k\rho_1\sigma_{E1}\sigma_{C1}\right) + \left(k^2\sigma_{E2}^2 + \sigma_{C2}^2 - 2k\rho_2\sigma_{E2}\sigma_{C2}\right)\right]}{\left[k\Delta_{ER} - \Delta_{CR}\right]^2} \quad (8.2)$$

Table 8.1 explains what each of the terms in Equation 8.2 represents.

The denominator of the expression $K\Delta_{ER} - \Delta_{CR}$ expresses the cost-effectiveness argument in terms of the desired INB (Chapter 2). However, we use $\lambda = K$, the cost-effectiveness threshold, for the remainder of this chapter. Therefore, in reference to point (a) in the above list, the measure of effectiveness is a single numerical quantity which combines separate measures of clinical efficacy and costs to form a measure of the economic efficiency that is worth pursuing. For example, if an existing treatment compared with the current standard of care offers an INB of £5000 (measured in money terms), then we might want to ensure that the new treatment has a sample size which can show INB values >£5000. Sometimes, the measure of effectiveness might also consist of two separate measures, quality of life (QoL) and efficacy, to form QALY. Therefore, the term Δ_{ER} in $K\Delta_{ER} - \Delta_{CR}$ might consist of a mean value for QoL (such as mean EQ-5D) and a mean value for a clinical measure (often a primary end-point measure, such as survival in cancer trials). The two can be combined to give a composite measure, such as a mean QALY. Therefore, the objective might

TABLE 8.1

Summary and Explanation of Terms for Equation 8.2

Term	Definition
σ_{Ej}^2	Population variance for clinical effect measure in treatment group 1
σ_{Ej}^2	Population variance for clinical effect measure in treatment group 2
σ_{Cj}^2	Population variance for cost in treatment group 1
σ_{Cj}^2	Population variance for cost in treatment group 2
n	Sample size per group (multiply by 2 for the total sample size in a two-group trial)
ρ_j	Correlation coefficient between efficacy and cost for treatment group 1
ρ_j	Correlation coefficient between efficacy and cost for treatment group 2
$K = \lambda$	The cost-effectiveness threshold unit cost the healthcare provider is prepared to pay to obtain a unit increase in effectiveness (in the United Kingdom, this is £20,000 to £30,000)
Δ_{ER}	Mean difference between groups 1 and 2 in terms of (clinical) effects
Δ_{CR}	Mean difference between groups 1 and 2 in terms of costs
R	Patient allocation ratio between two treatment groups (e.g. in 1:1 allocation, value is 1)
z_α	Value associated with false positive rate (typically 1.96 for 5% level)
z_β	Value associated with false negative rate (typically 1.28 for 20% (80% power) level)

be to estimate the sample size which yields a specific INB based on mean effects and costs.

Example 8.1: Sample Size Calculation for a Two Group Trial to Detect an Incremental Net (Monetary) Benefit of at Least £3000

In a two-group trial (Treatment A vs. B) in a 1:1 allocation (r = 1), the following parameters are considered to calculate the sample size for showing an INB of at least £3000.

We assume that the mean cost of the new experimental treatment (which happens to be an anti-cancer treatment for lung cancer) averages to £20,000 per year (Treatment A). The mean cost for Treatment B, the current standard of care, is £10,000, which is substantially cheaper. The new treatment will need to show a positive INB, and given that the costs are greater, needs to show how this additional (incremental) cost can be justified – what added value does the new treatment bring in terms of clinical effects (and QoL) to warrant paying the extra £47,000?

The clinical effects in this (hypothetical) example are based on mean estimates of QALYs which use QoL data from early phase trials and mean survival time data from Phase II trials. In the experimental arm, the QoL based on EQ-5D is a mean of 0.6 in each group. The mean survival (in economic evaluation, the mean survival time, not the median, is the measure of inter-est) for the new treatment was 17 months; in the standard treatment group, it is considered to be 14 months (so the mean survival time would have to be estimated through modelling or other methods). The QALY is therefore about 10.2 (0.6×17) for Treatment A and 9.4 (0.67×14) for Treatment B. If the CE threshold is £30,000, the number of patients required to show a posi-tive INB (e.g. £5000) on average, assuming no correlation between costs and effects ($\rho_1 = \rho_2 = 0$), can be calculated.

We assume that the variance of costs in the new and standard treatments is £10,000 and £6,000, respectively, and the variance of QALYs is 10.2 and 8.4 for Treatments A and B, respectively. The value of $\lambda = K = £30,000$. The denominator of Equation 8.2 is ($\lambda \times \Delta_E - \Delta_C)^2$: £30,000 × (10.2 – 9.4) – (£20,000 – £10,000) = £24,000 – £10,000 = $(£14,000)^2$.

Compared with the value of the INB under the null hypothesis (H_0: INB ≤ 0 vs. H_1: INB > 0), the value of £14,000 tells us that the new treatment is likely to offer (economic) benefit (i.e. INB > 0). Whether £14,000 per year is accept-able to the (tax)payer is a separate question. If the new treatment saves the payer, on average, £14,000 per year, the cumulative total across all patients treated may be significant. The sample size for each treatment group using Equation 8.2 reduces to the following:

$$n = (1.96 + 1.28)^2 \times \frac{\left[1 \times \left(£30,000^2 \times 3.5 + £10,000\right) + \left(£30,000^2 \times 2.9 + £6000\right)\right]}{\left[1 \times £30,000 \times (10.2 - 9.4) - (£20,000 - £10,000)\right]^2}$$

$$n = 605$$

Therefore, a sample size of 605 per group (1210 in total, allocated in a ratio of 1:1) will:

1. Show an INB > 0.
2. Reject H_0: that the new treatment would offer no value to the (tax) payer (INB < 0).
3. Be unlikely to have sufficient power to show INB values <£14,000. From a payer perspective, the larger the INB, the better.

Example 8.2: Taking into Account the Correlation between Cost and Effects

Now, assume that there is a correlation (which is equal in the two treat-ment groups) such that $\rho_1 = \rho_2 = 0.5$. That is, costs increase with effects. The correlations can be obtained from patient-level data in earlier tri-als, or from published literature or simulation. The correlations here are within each treatment group. In this case, entering $\rho_1 = \rho_2 = -0.9$ into Equation 8.2, assuming everything else is unchanged, gives

$$n = (1.96 + 1.28)2 \times \left[1 \times \left(£30{,}000^2 \times 3.5 + £10{,}000 - \left(2 \times 0.9 \times 1.73 \times 100 \times £30{,}000 \right) \right) \right.$$

$$\left. + \left(30{,}000^2 \times 2.9 + £6000 - \left(2 \times 0.9 \times 1.70 \times 77.45 \times £30{,}000 \right) \right) \right] /$$

$$1 \times £30{,}000 \times (10.2 - 9.4) - (£20{,}000 - £10{,}000)^2$$

$$n = 608 \text{ per group } (1216 \text{ in total})$$

Here we observe that sample size has increased (very slightly) due to the high negative correlation. At cost-effectiveness thresholds around £30,000, the impact of the correlation between costs and effects is negligible (see Figure 8.1). Glick (2011) notes that new health technologies with good outcomes are associated with lower cost and have a negative or win/win (lose/lose) correlation; therapies for asthma may provide examples of this correlation. On the other hand, therapies in which a good outcome is associated with a higher cost may provide examples of a positive or win/lose correlation, for example life-saving care.

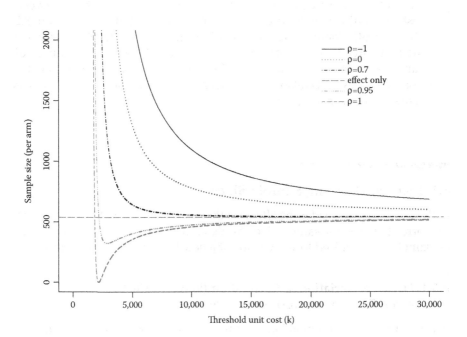

FIGURE 8.1
Sample size determination is plotted against the threshold unit cost, k, for different values of the correlation coefficient between cost and efficacy.

Example 8.3: Sample Size Calculation Assuming Negative Correlation between Costs and Effects

Now, assume that there is negative correlation (which is also equal in the two treatment groups) such that $\rho_1 = \rho_2 = -0.5$. That is, the costs increase with effects. The correlations can be obtained from patient-level data in earlier trials, or from published literature (although these are rarely, if ever, provided, so it is often better to make informed guesses). In this case, by entering $\rho_1 = \rho_2 = -0.5$ into Equation 8.2, assuming everything else is the same, the sample size can be determined by the reader as an exercise

(Hint: $n = (1.96 + 1.28)^2 \times [1 \times (\text{£}30{,}000^2 \times 3.5 + \text{£}10{,}000 + 2 \times 0.5 \times 1.73 \times 100 \times \text{£}30{,}000) + (30{,}000^2 \times 2.9 + \text{£}6000 + 2 \times 1.70 \times 77.45 \times \text{£}30{,}000)]/1 \times [\text{£}30{,}000 \times (10.2 - 7) - (\text{£}20{,}000 - \text{£}10{,}000)]^2)$.

Briggs & Gray (1998) present a conservative formula based on the ICER (Walker et al., 1995; Willan & O'Brien, 1996; Laska et al., 1999). This implicitly assumes a perfect negative correlation between cost and efficacy (i.e. as efficacy improves, costs will fall). Therefore, their formula resulted in a larger sample size to demonstrate cost-effectiveness than is required in practice. Later, Briggs & Tambour (2001) presented a more general formula using the INB approach, similar to the one we present. However, instead of considering a correlation coefficient between the cost and efficacy measures at the patient level for each treatment group, they consider a correlation between the cost and efficacy *differences*. The formula presented in Equation 8.2 is mathematically identical to that of Briggs & Tambour (2001) only if we assume that three components are equal: the correlation, the variances of costs and the variances of effects (see Technical Appendices 8B and 8C for proof of this). Glick (2011) also suggests using Fieller's theorem (Chapter 2) to estimate sample sizes for given inputs.

8.5 Factors Affecting Sample Sizes

In this section, we consider the effects of the cost-effectiveness threshold λ and the correlation ρ on the sample size. We also consider the concept of the smallest sample size required to show value of a new treatment in terms of INB.

8.5.1 Effects of Variation in Correlation (between Cost and Efficacy Measures) on Sample Sizes

Suppose that the correlation between the cost and efficacy in two treatment arms is equal ($\rho_1 = \rho_2 = \rho$). Now, under equal correlation, considering the sample size formula (Equation 8.2) as an identity and differentiating with respect to the correlation coefficient gives

$$\frac{\partial}{\partial \rho}(n_{CE}) = -2(z_\alpha + z_\beta)^2 \frac{k(\sigma_{E1}\sigma_{C1} + \sigma_{E2}\sigma_{C2})}{[k\Delta_{ER} - \Delta_{CR}]^2} \tag{8.3}$$

Since σ_{E1}, σ_{E2}, σ_{C1} and σ_{C2} are all positive, the partial derivative in Equation 8.3 is always negative for $\lambda > 0$, and the sample size is, therefore, a decreasing function of the correlation coefficient. That is, as the correlation increases, the sample size required to demonstrate value in terms of INB decreases. Briggs & Tambour (2001) also show that sample size is a decreasing function of the correlation coefficient based on their sample size formula. In Section 8.4, Table 8.2 shows how sample size varies for different values of the correlation coefficient: all else being equal, the more positive the correlation, the smaller the sample size needed; the more negative the correlation, the larger the sample size.

8.5.2 The Effect of the Cost-Effectiveness Threshold (λ) on the Sample Size Depends on the Correlation Coefficient

Under the assumptions that $\sigma_{E1} = \sigma_{E2} = \sigma_E$, $\sigma_{C1} = \sigma_{C2} = \sigma_C$ (i.e. the variances of costs and effects are equal within each treatment group) and $\rho_1 = \rho_2 = \rho$ are equal (i.e. the correlation between cost and effects in one treatment group is the same as in the other group), it can be shown that

$$n_{CE} \propto \frac{[k^2\sigma_E^2 + \sigma_C^2 - 2k\rho\sigma_E\sigma_C]}{[k\Delta_{ER} - \Delta_{CR}]^2} \tag{8.4}$$

When there is perfect negative correlation ($\rho = -1$), this expression becomes

$$\sqrt{n_{CE}} \propto \left[\frac{k\sigma_E + \sigma_C}{k\Delta_{ER} - \Delta_{CR}}\right] \tag{8.5}$$

TABLE 8.2

Sample Sizes for $\Delta_c = 1200$, $\delta = 0.7$ and $\omega = 0.975$ Using Frequentist Method

Effect (Correlation)	$\lambda = 10,000$	$\lambda = 20,000$	$\lambda = 30,000$
$\Delta_e = 0.8$ ($\rho_1 = \rho_2 = 0$)	912 (456/arm)	746 (373/arm)	702 (351/arm)
$\Delta_e = 1.5$ ($\rho_1 = \rho_2 = -0.5$)	268 (134/arm)	218 (109/arm)	204 (102/arm)
$\Delta_e = 1.5$ ($\rho_1 = \rho_2 = 0$)	222 (111/arm)	198 (99/arm)	192 (96/arm)
$\Delta_e = 1.5$ ($\rho_1 = \rho_2 = 0.5$)	176 (88/arm)	176 (88/arm)	178 (89/arm)

Provided that $k\Delta_{ER} - \Delta_{CR} > 0$ (INB >0), Equation 8.5 shows that the sample size decreases as the cost-effectiveness threshold increases. This makes sense intuitively, because if there are no limits on the affordability of a new treatment for the payer, then very little evidence might be needed to show cost-effectiveness. However, the sample size (n_{CE}) increases when the variances for costs and effects, σ_E and σ_C, increase. On the other hand, when there is perfect positive correlation ($\rho = 1$),

$$n_{CE} \propto \left[\frac{k\sigma_E - \sigma_C}{k\Delta_{ER} - \Delta_{CR}} \right]^2 \tag{8.6}$$

it can be shown that the sample size is an increasing function of λ for $k > \sigma_C / \sigma_E$ and a decreasing function of λ for $(\Delta_{CR}/\Delta_{ER}) < k > (\sigma_E/\sigma_C)$.

Therefore, for a very high positive correlation, the sample size can either increase or decrease with λ, depending on whether the value of λ is higher or lower than the variance ratio σ_C/σ_E. Therefore, the value of λ for which the sample size is minimum is

$$k_{min} = \frac{\sigma_C}{\sigma_E} \tag{8.7}$$

Figure 8.1 uses the data presented by Briggs & Tambour (2001). It clearly shows that sample size is a decreasing function of the correlation coefficient.

Briggs & Tambour display Figure 8.1 for $k \geq \sigma_C/\sigma_E$ instead of plotting for $k \geq \Delta_{CR}/\Delta_{ER}$, and our conclusions differ from their (incorrect) conclusion that a very strong positive correlation between the cost and efficacy differences may actually cause the sample size to be an increasing function of λ. Actually, for very high positive correlation, sample size is not a monotone function of λ; rather, it decreases (for $\Delta_{CR}/\Delta_{ER}) < k < (\sigma_C/\sigma_E)$) and then increases (for $k > \sigma_C/\sigma_E$) with increasing λ (for $\theta_R > 0$). For low correlation, it is a monotonically decreasing function of λ (for $\theta_R > 0$).

8.6 The Minimum Sample Size to Establish Cost-Effectiveness

This section now asks what is the smallest sample size required for a new treatment to show value in terms of the INB for a given cost-effectiveness threshold (λ). Taking the sample size formula from Equation 8.2 as an identity with equal correlation in both arms ($\rho_1 = \rho_2 = \rho$), differentiating with respect to λ, setting the partial derivative to zero and rearranging on λ gives

$$\lambda_{min} = \frac{\dfrac{\Delta_{CR}}{\Delta_{ER}}(\sigma_{E1}\sigma_{C1} + \sigma_{E2}\sigma_{C2})\rho - (\sigma_{C1}^2 + \sigma_{C2}^2)}{\dfrac{\Delta_{CR}}{\Delta_{ER}}(\sigma_{E1}^2 + \sigma_{E2}^2) - \rho(\sigma_{E1}\sigma_{C1} + \sigma_{E2}\sigma_{C2})} \tag{8.8}$$

Here, λ_{min} denotes the value of λ for which the sample size is the smallest to show cost-effectiveness. Equation 8.8 indicates the existence of a minimum sample size depending on the correlation coefficient. It is apparent from Equation 8.7 that for $\rho \le 0$, the value of λ_{min} is negative, which is not useful in practice (because the cost-effectiveness threshold should be >0).

Equation 8.8 shows that when

$$\rho = \left[\Delta_{CR}/\Delta_{ER}\right]\Big/\left[(\sigma_{E1}\sigma_{C1} + \sigma_{E2}\sigma_{C2})\big/(\sigma_{E1}^2 + \sigma_{E2}^2)\right]$$

then $k_{min} \to \infty$.

For

$$\rho > \left[\Delta_{CR}/\Delta_{ER}\right]\Big/\left[(\sigma_{E1}\sigma_{C1} + \sigma_{E2}\sigma_{C2})\big/(\sigma_{E1}^2 + \sigma_{E2}^2)\right]$$

the minimum sample size exists at finite $k_{min} > [\Delta_{CR}/\Delta_{ER}]$. Therefore, the critical point of correlation is

$$\rho = \left[\Delta_{CR}/\Delta_{ER}\right]\Big/\left[(\sigma_{E1}\sigma_{C1} + \sigma_{E2}\sigma_{C2})\big/(\sigma_{E1}^2 + \sigma_{E2}^2)\right] \tag{8.9}$$

For the special case of $\sigma_{E1} = \sigma_{E2} = \sigma_E$, $\sigma_{C1} = \sigma_{C2} = \sigma_C$, Equations 8.8 and 8.9 become, respectively,

$$k_{min} = \frac{\sigma_C}{\sigma_E}\left[\rho\frac{\Delta_{CR}}{\Delta_{ER}} - \frac{\sigma_C}{\sigma_E}\right]\Big/\left[\frac{\Delta_{CR}}{\Delta_{ER}} - \rho\frac{\sigma_C}{\sigma_E}\right] \tag{8.10}$$

and

$$\rho = \frac{\Delta_{CR}}{\Delta_{ER}}\Big/\frac{\sigma_C}{\sigma_E} \tag{8.11}$$

For the perfect positive correlation ($\rho = 1$), Equation 8.11 reduces to Equation 8.2.

8.7 Bayesian Sample Size Approach

O'Hagan & Stevens (2001) proposed a more general sample size formula for cost-effectiveness analysis from a Bayesian perspective. They consider a correlation between the measures of efficacy and cost at the patient level in each arm of the trial. They state two objectives. The first, referred to as the analysis objective, is that the outcome of the trial will be considered positive if the posterior probability that INB is positive (i.e. the new treatment is cost-effective) is at least $1-\alpha$. The second, referred to as the design objective, is that the sample size should be sufficiently large that the probability of achieving a positive result is at least $1-\beta$. Unlike the most natural Bayesian analysis (a single prior is used without reference to analysis and design stage), they proposed a novel solution based on assuming two prior distributions: one for the design stage and one for the analysis stage. Under the very strong design prior and weak (improper) analysis prior, their sample size formula provides exactly the same solution as the frequentist solution of Equation 8.2. Whitehead et al. (2008) also state that for certain 'non-informative priors', Bayesian sample size determination coincides with the corresponding frequentist formula.

Let us assume that the prior mean and variance are m_d and v_d at the design stage and m_a and v_a at the analysis stage. O'Hagan & Stevens (2001) show that the analysis objective leads to the following condition for a positive outcome:

$$\sqrt{v_1}\left(v_a^{-1}m_a + n\sigma_+^{-2}\hat{\theta}\right) \geq z_\alpha$$

where

$$\sigma_+^2 = k^2\left(\sigma_{E1}^2 + \sigma_{E2}^2\right) + \left(\sigma_{C1}^2 + \sigma_{C2}^2\right) - 2k\left(\rho_1\sigma_{E1}\sigma_{C1} + \rho_2\sigma_{E2}\sigma_{C2}\right)$$

$\hat{\theta}$ is the estimate of θ from the trial

$$v_1 = \left(v_a^{-1} + n\sigma_+^{-2}\right)^{-1}$$

They also show that to fulfil the design objective, sample size n must be large enough to satisfy the following condition:

$$\left(v_a^{-1}m_a + n\sigma_+^{-2}m_d\right) - \left\{z_\alpha / \sqrt{v_1}\right\} - z_\beta\sqrt{\left(n\sigma_+^{-2}\right)^2\left(v_d + n^{-1}\sigma_+^{-2}\right)} \geq 0 \tag{8.12}$$

This inequality has a finite solution if, and only if, $\Phi\left(m_d/\sqrt{v_d}\right) > 1-\beta$. The most useful case in practice is the case in which we have informative design prior but the analysis prior is weak. For situations in which $v_a^{-1} = 0$

(i.e. analysis prior is weak) and $v_d = 0$ (i.e. design prior is strong), we have the frequentist's solution, and this inequality reduces to $\hat{\theta} \geq z_\alpha \sigma_+ / \sqrt{n}$ with the solution $n = \{(z_\alpha + z_\beta)\sigma_+ / m_d\}^2$. However, if this variance is non-zero, we will have a larger sample size than we can expect in the similar situation for the frequentist approach.

8.7.1 The Effect of the Correlation and λ on the Bayesian Sample Size

There exists a similar relationship between λ and n, for different values of a correlation coefficient between the measures of efficacy and cost, as in the frequentist approach. In the case of a very high positive correlation between the measures of efficacy and cost, the relationship between λ and n can be non-monotone, as in the frequentist approach.

Although the cost-effectiveness methodologies have been prevalent for almost two decades, sample size calculations for cost-effectiveness trials are still limited in the applied area of clinical trials due to lack of software. A C+ program is available (Sarker et al., 2012), and a Windows-based 'click and push' application, SCET-VA® version 1.1 (Khan, 2015), which computes sample sizes under both frequentist and Bayesian frameworks, is available from the author on request. The software may be requested by contacting the author at https://www.researchgate.net/profile/Iftekhar_Khan5.

> **Example 8.4: Using a Weak Analysis Prior and a Weak Design Prior**
>
> We use the same example data as were used by O'Hagan & Stevens (2001). They assume that the standard deviations for the effects and costs are $\sigma_{E1} = \sigma_{E2} = 4.04$ and $\sigma_{C1} = \sigma_{C2} = 8700$, respectively. In this case, the analysis prior expectation and design prior expectations are assumed to be equal (i.e. $m_a = m_d = [5, £6000, 6.5, £7200]$). This implies that the treatment effect is 0.8 and the difference in costs between the two treatments is £1200. The analysis prior variance is also assumed to be a weak (non-informative) prior (i.e. $v_a = 0$), and the design prior variance is also assumed to be weak, such that $v_a = v_d = 0$. Hence, with $\Delta_c = 1200$, $\Delta_e = 0.8$, $\delta = 1 - \beta = 0.7$, $\omega = 1 - \alpha = 0.975$. This shows that the sample size required to demonstrate cost-effectiveness under a cost-effectiveness threshold of £10,000 with at least 70% assurance is 456 per arm (912 in total). Table 8.2 shows the sample sizes generated by Sarker et al. (2012) for two different effect sizes (assuming the costs are the same) and three different correlation coefficient values based on the frequentist sample size formula in Equation 8.12.

Using the sample size software SCET-VA® V1.1, the output is shown in the following screenshots for the data in Table 8.2, row 3, column 1 ($\lambda = 10,000$, $\Delta_e = 1.5$, $\rho_1 = \rho_2 = 0$):

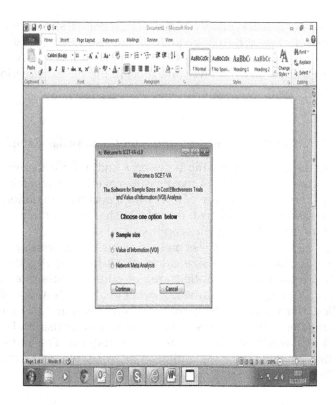

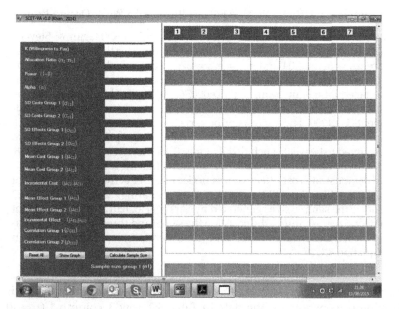

This is an example output of the software.

In Figure 8.1 (Section 8.3), we showed sample size as a decreasing function of the correlation coefficient. The results of Table 8.2 confirm this. However, the rate of decrease falls as the CE threshold increases (Table 8.2). Also, the sample size decreases with increasing values of λ. This means that if $\lambda \to \infty$, the sample size will reduce to the sample size based on efficacy only.

Example 8.5: Bayesian Sample Size Using a Weak Analysis Prior and a Proper Design Prior

In this example, we assume now that the prior analysis and design expectations are in the same form as in Example 8.1, that is, $\mathbf{m}_a = \mathbf{m}_d = [5, £6000, 6.5, £7200]$. We also assume that the prior analysis variance $\mathbf{v}_a$ is still a weak prior set equal to 0, but the design prior variance $\mathbf{v}_d$ is proper and takes the form

$$\mathbf{v}_d = \begin{pmatrix} 4 & 0 & 3 & 0 \\ 0 & 10^7 & 0 & 0 \\ 3 & 0 & 4 & 0 \\ 0 & 0 & 0 & 10^7 \end{pmatrix}$$

Hence, with $\Delta_c = 1{,}200$, $\Delta_e > 0.8$, and varying λ from 10,000 to 30,000 (the cost-effectiveness threshold), $\delta = 0.7$, $\omega = 0.975$, $\mathbf{v}_a = 0$, and the correlation between cost and effect $\rho_1 = \rho_2 = 0$, the sample size required to demonstrate cost-effectiveness is shown in Table 8.3. Clearly, the sample size is larger compared with the frequentist scenario.

TABLE 8.3

Sample Sizes for $\Delta_c = 1200$, $\delta = 0.7$ and $\omega = 0.975$ Using a Proper Prior at Design Stage and a Weak Prior at Analysis Stage

Effect	$\lambda = 10{,}000$	$\lambda = 20{,}000$	$\lambda = 30{,}000$
$\Delta_e = 1.5$	764 (382/arm)	570 (285/arm)	528 (264/arm)
$\Delta_e = 1.0$	25,368 (12,684/arm)	7,218 (3,609/arm)	5,604 (2,802/arm)

Note: Sample sizes for effect sizes ≤0.8 are very large and unlikely to be used in practice for the chosen values of δ and ω, and are therefore not included in Table 8.2.

Example 8.6: Bayesian Sample Size Where the Analysis Prior Is Set Equal to the Design Prior Used in Example 8.4

In this example, we generate sample sizes using proper priors at both the design and analysis stages (see Table 8.4). As expected, the sample size increases further when we have proper prior at both the design and analysis stages.

Example 8.7: From a Lung Cancer Published Study

In this example, we calculate the sample size retrospectively, assuming a weak analysis prior expectation and variance and a proper prior variance matrix for the design stage.

In this lung cancer study comparing Erlotinib with Placebo (Bradbury et al., 2010), cost-effectiveness was also a key end point. We provide a retrospective power calculation for various values of λ ranging between £20,000 and £30,000 (the current NICE cost-effectiveness threshold), $\delta = 0.7$ and $\omega = 0.975$. However, because the published costs are presented in Canadian dollars, we provide values of λ in Canadian dollars, ranging from about $16,700 to $50,000 for λ values from £10,000 to £30,000 in Table 8.5 (approximate current exchange rate of Can$1 = £0.62). The total sample size used in this study was 731 (487 vs. 244 in a 2:1 ratio) to demonstrate a clinically important difference in terms of survival benefit.

TABLE 8.4

Sample Sizes for $\Delta_c = 1200$, $\delta = 0.7$ and $\omega = 0.975$ Using Proper Priors at both Design and Analysis stage

Effect	$\lambda = 10{,}000$	$\lambda = 20{,}000$	$\lambda = 30{,}000$
$\Delta_e = 1.5$	804 (402/arm)	602 (301/arm)	560 (280/arm)
$\Delta_e = 1.0$	25,466 (12,773/arm)	7,264 (3,632/arm)	5,644 (2,822/arm)

Note that sample sizes for effect sizes ≤0.8 are very large and unlikely to be used in practice for the chosen values of δ and ω, and therefore are not included in Table 8.3.

TABLE 8.5

Sample Sizes for $\Delta_c = \$12{,}301$, $\delta = 0.7$ and $\omega = 0.975$ Using Proper Prior at the Design Stage Only and Weak Analysis Prior

Effect	$\lambda = \$16{,}700$	$\lambda = \$33{,}300$	$\lambda = \$50{,}000$
$\Delta_e = 1.6$	3,146 (1,573/arm)	608 (304/arm)	436 (218/arm) 222 (111/arm)[a]

[a] Frequentist solution requires sample size of 222 (111/arm).

The observed mean costs for Erlotinib and Placebo were $16,485 and $4,184, respectively from Table 8.2 of Briggs & Tambour (2001). Although no information was provided on the variability of these estimates, Table 8.2 of Briggs & Tambour (2001) did provide estimates of variability for the cost subcomponents. In any case, following O'Hagan & Stevens (2001), we assume a weak prior variance for the costs at both analysis and design stages, because less information would be known about costs in practice. In addition, the mean clinical effect (overall survival [OS]) was 9 months for Erlotinib (SD = 0.75) and 7.4 months (SD = 0.62) for Placebo. Recall that in economic evaluation, the mean rather than the median is the statistic used for survival end points (in this example, the median OS is 6.7 and 4.7 months for Erlotinib and Placebo, respectively). We therefore assume $\mathbf{m}_a = \mathbf{m}_d = [9.0, \$16{,}485, 7.4, \$4{,}184]$.

For the prior distribution at the analysis stage, we assume the matrix, $\mathbf{v}_a = 0$. However, following Example 8.4, we assume a proper design prior variance, $\mathbf{v}_d$. One difference here is that we also set the covariance equal to 0 (previously the covariance was set equal to 3 in Example 8.4):

$$\mathbf{v}_d = \begin{pmatrix} 0.75^2 & 0 & 0 & 0 \\ 0 & 10^7 & 0 & 0 \\ 0 & 0 & 0.75^2 & 0 \\ 0 & 0 & 0 & 10^7 \end{pmatrix}$$

With the correlation between costs and effects assumed to be $\rho_1 = \rho_2 = 0$, the sample size required to demonstrate cost-effectiveness with at least 70% assurance and protection against an incorrect decision in concluding that Erlotinib is cost-effective (when in reality it is not) set at 2.5%, the required sample size would be 436 (218 per arm) to demonstrate cost-effectiveness at a threshold of $50,000. Table 8.5 shows the sample sizes for proper design priors and varying λ in Canadian dollars. For the same threshold of $50,000, the frequentist sample size, assuming weak analysis and design priors, would be 222 patients in total.

8.8 The Normality Assumption

Cost data are always positive and often exhibit skewed distributions. However, the normality assumption is needed at the patient level for both the frequentist and the Bayesian approach. The normality assumption of the sample means is justified in the frequentist approach for large samples by virtue of the central limit theorem. O'Hagan & Stevens (2002) suggest that the approximation is generally very good for sample sizes of 50 or more. In practice, the greater the skewness of the cost data, the larger the sample size required for the normality assumption to have validity. Although this normality assumption is technically valid, there is a concern about its applicability when there is a particularly acute skewness, or sample sizes are not large. However, the majority of trials in which a cost-effectiveness analysis is required with patient-level data are larger Phase III trials, and this is unlikely to be a great concern. According to O'Hagan & Stevens (2001), 'Even if normality of sample means is reasonable, when the original data are not normal the sample means will generally not provide the most efficient inferences'. O'Hagan & Stevens (2002) advise that cost-effectiveness analysis should be based on careful modelling (see also Chapter 4) of the underlying distributions, rather than relying on distribution-free methods such as the non-parametric bootstrap or a normal theory based on the central limit theorem. According to Nixon & Thompson (2004), 'skewed parametric distributions fit cost data better than the normal distribution, and should in principle be preferred for estimating the population mean cost'. The main motivation for considering skewed parametric distributions is to improve the precision of the estimated population mean cost.

Assuming that the cost data are distributed as log-normal while efficacy data are normal, Sarker & Whitehead (2004) rewrote their frequentist sample size formula as follows:

$$n_{CE} = 2(z_\alpha + z_\beta)^2 \frac{\left[k^2\sigma_E^2 + \sigma_C^{*2} - 2k\rho^* \sigma_E \sigma_C^*\right]}{\left[k\Delta_{ER} - \Delta_C^*\right]^2}$$

where:
λ is based on log(cost), instead of the original cost scale
ρ^* is the correlation coefficient between the measures of efficacy and log(cost)

In sample size calculations, differences in mean costs are hypothesised values, whereas values of variance quantities and the correlation coefficient are assumed to be known; an assumption that data are log-normally distributed would not necessarily change the required sample size. This assumption may, however, improve the precision of the estimated population mean cost. For other skewed parametric distributions of cost (e.g. gamma and log-logistic), derivations of the sample size formula would be difficult.

In the Bayesian context, the sample size formula based on the log-normal cost is again similar to the formula based on a bivariate normal assumption, where λ and variance components for the cost data are defined in terms of log(cost), and the correlation coefficient is defined between the measures of efficacy and log(cost).

8.9 Obtaining the Necessary Data and Tools for Calculating Sample Size

Glick (2011) discusses where to obtain the necessary data for the calculation of sample size for a cost-effectiveness trial. In general, estimates of mean cost and effects can be obtained from earlier trial data. Individual variabilities of effects for QoL can be obtained from mapping functions (see Chapter 5), because these allow patient-level estimates of utilities. Some Phase I trials can offer ideas on adverse events (usually patient-based Phase I trials). However, care should be taken, because Phase I trials are also conducted in healthy volunteers, and the target patient population may be very different: for example, Phase I trials in very sick patients who try several treatments do not accurately reflect a fair cost comparison in similar patients who take only one treatment.

Glick (2010) provides some STATA programs for frequentist cost-effectiveness sample sizes, which are freely available at http://www.uphs.upenn. edu/dgimhsr/stat-samps.htm. Other software in a beta version is available from http://www.healthstrategy.com/cgi-bin/samplesize.pl (Health Decision Strategies, LLC, Princeton, New Jersey, United States). New software SCET-VA® V1.1 (Khan, 2015), based on Sarker et al. (2013), allows Bayesian and frequentist sample sizes to be calculated with a simple 'click and push' application. The software can be used to calculate sample sizes using VOI methods as well as a module for mixed treatment comparisons (Section 8.7.1). An example of the software is shown below and was demonstrated in a Bayesian context earlier.

In summary, sample sizes for cost-effectiveness methods are still relatively new. Views may vary from country to country as to whether one can offer a hypothesis-testing framework for demonstrating value of new treatments. To say that a new healthcare technology has demonstrated value (using INB) such that the chance of an incorrect conclusion on value is limited to 5% is still a strong statement if this can be backed by clinical and economic data. The Bayesian approach also offers an equal (if not more powerful) statement when the INB is >0. Sample sizes that enable making these statements are therefore likely to become important in making the case for value of new treatments. VOI methods using Bayesian techniques are alternative ways for estimating the sample sizes.

8.10 Value of Information

In Chapter 6 we saw how cost-effectiveness models were used to estimate the INB and the ICER. In addition, we discussed how the model inputs could be subject to uncertainty analysis using one-way, two-way or probabilistic sensitivity analysis (PSA). In Chapter 8, it was shown how the cost-effectiveness acceptability curve (CEAC) was generated from PSA; VOI can be considered as an extension of PSA. However, whereas the CEAC yields estimates of the probability of cost-effectiveness of the new treatment, VOI allows us to quantify the consequences of selecting the wrong treatment. Recall that the expected net benefit (ENB) was

$$\text{Effects} \times \lambda - \text{Costs}$$

where λ is the cost-effectiveness threshold and Effects are measures of effectiveness (e.g. QALYs).

At the time of a cost-effectiveness analysis, and during a review of the evidence for the cost-effectiveness of a new treatment by a reimbursement agency, any decision to reimburse the treatment(s) based on the highest ENB is still subject to uncertainty (i.e. the payer may select the wrong treatment to reimburse). Since uncertainty almost always exists, the chance of a wrong decision is always possible. With a wrong decision, however, there is always a cost or consequence. For example, an economic evaluation might be performed which results in the new treatment yielding an INB >0. However, at the time of marketing authorisation, the pharmaceutical company might be requested to carry out a post-marketing commitment study to evaluate the longer-term effects (e.g. side effects) of the new treatment. Consequently, from the post-marketing study, it transpires that the adverse event rates were slightly higher for the experimental/new group and slightly higher for the control/standard group. When the INB was recalculated (after the post-marketing study), the INB was not as large as initially estimated, leading to a larger ICER than originally reported. A decision at the time of the first analysis to declare a new treatment as being cost-effective might have been considered too risky from a payer perspective if the information that is available now was available then.

Quantifying the chance of uncertainty can be assessed by the probability that a decision based on the ENB is incorrect. The magnitude of the opportunity loss (in ENB) due to the uncertainty can be quantified and is known as the *expected value of perfect information* (EVPI), because it is the amount of ENB that a decision maker would need to pay to remove all the uncertainty. More formally, the EVPI is the amount by which the ENB changes (upwards or downwards) if the uncertainty of all parameters (e.g. costs, effects, adverse event rates) used in a decision to choose a particular treatment for reimbursement is simultaneously eliminated. In some cases, additional information (through more research) might be sought on some specific parameters (e.g. just efficacy end points). Specific parameters might be those whose uncertainty is largest (because there was very little data about them before the trial started), and the value (in terms of changes to the ENB) of reducing this will be estimated. This is called the *expected value of partial perfect information* (EVPPI).

However, in order to have 'perfect information' (i.e. to eliminate all sources of uncertainty), we would need a very large sample size. Hence, 'perfect information' is an ideal (just as having an unlimited sample size is in a clinical trial) to answer a specific question. Therefore, in practice we try to quantify the uncertainty of decisions (of choosing treatments) based on sample sizes available. As such, we are interested in the *expected value of sample information* (EVSI). The EVSI compares the ENB based on a trial given a current sample size (imperfect information based on current data) with the ENB if more data were collected (approaching perfect information – since the greater the sample size, the closer to perfect information one gets). Hence, the EVSI is considered to be another approach to estimating the optimal sample size for a future study. The question becomes, for example, 'How much value (based on ENB) is there in having more data (i.e. a larger sample size than the one we currently have) so that the current decision does (or does not) change?' We now show an example of how the EVSI can be calculated and used in the context of sample size calculation. The EVSI is a fully Bayesian approach.

Example 8.8: Simple Description of EVSI

We adapt an example from Ades et al. (2004), following their algorithms to compute the EVSI.

BACKGROUND

A clinical trial has been performed (n = 600; 300 per treatment group). The experimental treatment group (Treatment A) was shown to be cost-effective over Treatment B (standard treatment) with a current ENB of £30,000 for Treatment A and £20,000 for Treatment B. However, the reimbursement agency stated that data were needed to better estimate the uncertainty of long-term toxicity data (which can have a significant cost burden). Although the initial decision looks favourable for Treatment A (i.e. a higher ENB), the payer review groups have suggested that the sponsor continue to investigate whether adverse event rates (perhaps using a two-arm observational study) might change any decision made now. One question is what sample size should be used so that a (currently favourable) decision remains favourable, or whether certain design parameters need to be considered to avoid a payer switching decision (from Treatment A to Treatment B).

As a starting point, let us assume that the additional observational study is planned as a sample size of n = 200 (100 per group). The parameter of particular interest is the (long-term) adverse event rate, P_{AE}, estimated as 25% for Treatment A and 10% for B in the initial randomised controlled trial (RCT) (n = 600 patients). The question is whether additional data gathered in the form of an observational study impacts the ENBs. Currently, the decision is in favour of Treatment A ($ENB_A >$ $ENB_B = £30,000 - £29,000 = £1,000$), as shown in Table 8.6. The maximum ENB is £30,000 in favour of Treatment A.

Step 1:

First calculate the ENB for each treatment group.

Step 2:

Determine which parameters will be subject to uncertainty, and determine the distribution to be used. The main reason for a future observational study is to get a better estimate of the values of the parameter, P_{AE} (since only one RCT has ever been conducted in this indication). One concern is that, from current data, the current adverse event rate P_{AE} remains uncertain, and a further observational study with a sample size of n = 300 (150 per group) may help to reduce uncertainty of the decision. Therefore, P_{AE} will be simulated 10,000 times from a beta binomial (BB) distribution. The BB distribution has parameters (α, β), so we need to choose a set of parameters whose mean will result in about 0.25 for Treatment A and 0.10 for B, using the fact that

$$\alpha = \mu \times \left[\left((\mu \times (1-\mu))/\sigma^2 \right) - 1 \right] \tag{8.13}$$

$$\beta = (1-\mu) \times \left[\left(\mu \times (1-\mu) \right)/\sigma^2 - 1 \right] \tag{8.14}$$

In this example, using an estimate of $\alpha = 3$ and $\beta = 9$ for Treatment A yields an estimate with mean = 0.25 (using mean = $\alpha/\alpha + \beta$, from Equation 8.13). Simulation from a BB is carried out for each treatment group separately.

TABLE 8.6

Model Inputs for EVSI Computations

Treatment	P_{AE}	Cost of Treatment (£)	QALY	ENB[a] (£)
A	0.25	150,000	6	30,000
B	0.10	100,000	4.3	29,000

[a] $ENB = -Cost + \lambda \times Effect$: $ENB_A = -£150,000 + 6 \times £30,000$; $ENB_B = -£100,000 + 4.3 \times £30,000$ (assuming a cost-effectiveness threshold λ of £30,000).

Data from a BB can be simulated in SAS using the fact that the ratio of two gamma distributed variables is a beta:

```
Z1=rangam(seed,alpha);
Z2=rangam(seed,beta);
X=Z1/(Z1+Z2); **gamma (alpha, beta)
```

Step 3:

Once 10,000 values of P_{AE} for each of the treatments have been generated, for each simulated P_{AE}, we then generate data from binomial ($n = 150$, P_{AE}), each simulated data set being of sample size $n = 150$. The likelihood is therefore generated from a binomial ($n = 150$, P^*_{AE}).

Step 4:

The next step is to compute the posterior mean adverse event rate. The prior BB combined with a binomial likelihood yields a posterior mean of P^*_{AE} of the form

$$a*/a*+b*$$

Example 8.9: Beta Binomial Prior and Binomial Likelihood

Before Any Data Are Observed

Before any data were observed, θ (i.e. P_{AE} for Treatment A) was assumed to be a beta (2,2). Since the mean of a beta (a,b) = a/a + b, this corresponds to a prior mean of (2/2 + 2 = 50%).

The form of a prior beta (1,3) is

$$\frac{1}{\text{beta }(1,3)}\theta(1-\theta)$$

In general, a beta (a,b) density is

$$\frac{1}{\text{beta }(a,b)}\theta^{a-1}(1-\theta)^{b-1}.$$

After Data Are Observed

Now assume that 10 further patients are observed for any adverse events. These outcomes are assumed to be dichotomous (Yes/No type outcomes). Assume that the 10 subjects have the following observed data:

$$x_i = (1,0,1,0,0,0,0,1,0,0)$$

where x_i is an indicator variable for 1 if an AE is observed and 0 if not.

For this realisation of AEs, the P_{AE} rate can now be updated by computing the posterior mean. In order to do this, we need the likelihood of the 10 outcomes.

$$L(\theta;x) = \prod_{i=1}^{10} \theta^{x_i}(1-\theta)^{1-x_i} = \theta^3(1-\theta)^7$$

We now combine the prior distribution with the likelihood to form the posterior distribution of P_{AE}:

$$P(\theta;x) \propto \theta^3(1-\theta)^7 \theta(1-\theta) = \theta^4(1-\theta)^8$$

Compare this with

$$\frac{1}{Beta\,(a,b)}\ \theta^{a-1}(1-\theta)^{b-1}$$

(a posterior beta (5,9))
Here $a - 1 = 4$ and $b - 1 = 8$, giving $a = 5$ and $b = 9$.

Prior	Beta (2,2)	2/2 + 2 = 50%
Posterior	Beta (5,9)	5/5 + 9 = 36%

The proportionality symbol ($\propto$) means that the full computation of the likelihood with the prior is determined through integrating out the prior × likelihood such that some terms cancel out.

From the computations (integrations), this results in a posterior beta (5,9). The mean of a beta $(a,b) = a/(a + b)$. Therefore, the posterior (mean) P_{AE}, after having observed a further 10 outcomes, is $5/(5 + 9) = 36\%$. The new (additional) data have revised the estimate of P_{AE} from 50% to 36%.

Hence, we will generate 10,000 posterior mean values $P^*_{i\,AE}$, where the subscript i is from 1 to 10,000, for each simulated value. An example of the format of the simulated data is shown in Table 8.7.

Step 5:

For each simulated posterior mean P_{AE}, compute the ENB for each treatment group. That is, we would have 10,000 ENB values each for Treatments A and B.

Hence, by formulating a prior distribution based on the original $n = 600$ from the RCT (which is a large amount of data to have an informative prior), and further simulating $n = 300$ ($n = 150$ per treatment group if a two-group observational study), we now have 10,000 (updated) ENBs for each of the treatments. We are now in a position for each simulated ENB to compute the impact of the additional $n = 300$ patients through the EVSI.

Step 6:

For each simulated data set, we compute the maximum of the ENB between treatments. For example, if after the first simulation for the posterior mean of P_{AE}, the ENB is £37,000 for Treatment A and is £31,000 for

Treatment B, we will retain the greater of these values (£37,000). This maximum is then compared with (subtracted from) the ENB under the current decision, that is, £33,000, as shown in Table 8.7.

Table 8.7 shows how the entire data structure might look after one simulation.

Based on Table 8.7, after one cycle of a Monte-Carlo simulation, a decision maker (payer) may switch from considering the new treatment (Treatment A) having a greater ENB in favour of Treatment B because the ENB for Treatment B is greater than that under the current decision. The process continues for each cycle of the simulation, so that we generate a posterior distribution of differences.

Step 7:

For all the 10,000 differences between current ENB and ENB under the assumption that further data will be collected, we can compute the mean of all the differences (Table 8.7) to generate the EVSI, which is £6120. This was evaluated for a sample size of n = 150 per group; however, the exercise can be repeated for sample sizes over various ranges.

In addition to the EVSI, we can also calculate the number and proportion of times out of 10,000 that the ENB under the current decision is greater than under the assumptions of additional data (i.e. the chance of a switch in the payer decision). In this example, in about 33% of the 10,000 simulations, the ENB was in favour of the standard treatment B. This implies that the decision maker will make an error 33% of the time if the sample size is 300 for a future observational study. Further results are summarised in Table 8.8 using a format similar to that described in Ades et al. (2004). The software (SCET-VA-version 1.0®) can be used to perform these computations for the choice of two priors (normal and beta in this case).

TABLE 8.7

Example Realisation of Value of Sample Information after One Simulation of P^A_{AE} and P^B_{AE}, the Adverse Event Rates for Treatments A and B, respectively

Simulation Number	P^A_{AE} P^B_{AE}	Current Decision			Posterior Decision		VSI (Max)
		ENB_A	ENB_B	(Max)	ENB_A	ENB_B	
prior	25%	33,000	30,000	30,000	–		–37,000
	10%		33,000	33,000			–4,000[a]
1	28%	33,000			31,000	37,000	
	15%						
2	..	..	..		..		
..							
10,000							
Mean (EVSI)							**6,120**

Note: VSI = Value of sample information. EVSI = expected value, which is calculated from the average of all the 10,000 VSI values.

[a] Calculated as 33,000 – 37,000 = –4,000.

TABLE 8.8

EVSI and Chance That the ENB under the Plan of an
Additional Study Is > ENB under the Current Decision

Sample size (n$_i$)	EVSI (£)	Pr[ENB$_{n=ni}$] > Pr[ENB$_{current}$]
50	1,300	0.15
100	2,400	0.22
200	5,280	0.27
400	6,090	0.29
600	6,120	0.33
1,000	6,180	0.38
2,000	6,185	0.39
5,000	6,190	0.40
10,000	6,192	0.40

The chance of switching decision appears to plateau after 1000 patients
(Table 8.8). Hence, any further data collected by way of clinical trials or observational studies are likely to result in 'erroneous' decisions not greater than
40. One may perform EVSI computations by taking into account uncertainty
in QoL or other parameters. The choice of the prior would be important, and
it might be preferable to simulate from correlated prior distributions of costs
and effects, which may be technically challenging.

A separate question is whether 33% is an acceptable risk to the decision
maker. In comparison with a Type I error of 5% for marketing approval, this
value might be high. One (perhaps ethical) issue is whether patients should
continue to be treated with the new treatment even if there is a risk that the
payer might be wrong 33% of the time. The risk is not that the new treatment
is not efficacious, but, rather, that it is not cost-effective. This risk to the public
(declaring that the new treatment is efficacious when in reality it is not) set
at 5% (frequentist 5% Type I error) is unlikely to be given the same weight as
a wrong decision on economic grounds. Whether this is right or wrong is a
matter for debate.

Exercises for Chapter 8

1. It is not ethical to power a clinical trial for an economic evaluation.
   Discuss.

2. 'Sample size calculations for cost-effectiveness analyses are not
   practically useful because the clinical end point is more important.'
   Criticise this statement.

3. Using the software SCET-VA V 1.1 or otherwise (you can request this from the author at https://www.researchgate.net/profile/Iftekhar_Khan5), find the sample size for the following:

   a. Mean cost for Treatment A = £8000, SD = 400

   b. Mean cost for Treatment B = £7000, SD = 320

   c. Mean effect for Treatment A = 1.3 QALY, SD = 0.92

   d. Mean effect for Treatment B = 1.0, SD = 0.88

   Assume correlation between costs and effects is zero for each treatment group.

4. Using the same data as in Question 3, recalculate the sample size when there is a correlation of 0.8 between costs and effects in each group. Interpret your answer. Repeat for a correlation of −0.8; explain your result.

5. Using the software, confirm that the results of a frequentist and a Bayesian analysis are the same for the data in Question 3, assuming the analysis and design priors for the variance are weak (i.e. set to zero).

6. What is value of information analysis, and how can it be a useful analysis to present to a reimbursement authority?

7. Use the software to confirm (approximately) the results in Table A2 for N = 100 from Ades et al. (2004).

Appendix 8A: SAS/STATA Code

See software SCET-VA version 1.1 for sample size and VOI.

Technical Appendix 8B: Derivation of Sample Size Formula

Suppose that the study to be designed will comprise a clinical trial in which patients are randomised between two therapies: the comparator treatment (Treatment 1) and the experimental treatment (Treatment 2). Suppose that the data will yield a measure of efficacy and a cost for each patient. Let x_{Eji} and x_{Cji} be measures of effectiveness and cost on the ith patient of the jth treatment: $j = 1, 2; i = 1, 2,..., n_j$; and n_j is the respective sample size. The expectation of x_{Eji} is denoted by $E(x_{Eji}) = \mu_{Ej}$, the population mean efficacy under treatment j, and the expectation of x_{Cji} is denoted by $E(x_{Cji}) = \mu_{Cj}$, the population mean cost under treatment j. For the population variances, let $\text{Var}(x_{Eji}) = \sigma_{Ej}^2$ and $\text{Var}(x_{Cji}) = \sigma_{Cj}^2$. Since the sample size here is derived based on the net benefits (Chapter 2), which is a linear function of the mean cost and effectiveness rather than the ICER, the dependence between the cost and effectiveness measures is determined by a correlation coefficient: $\text{Corr}(x_{Eji}, x_{Cji}) = \rho_j$. Here, we assume that (x_{E1i}, x_{C1i}) and $(x_{E2i'}, x_{C2i'})$ are mutually independent over all i and i′ for $i \neq i'$.

Now the net (monetary) benefits can be written as

$$\theta = k(\mu_{E2} - \mu_{E1}) - (\mu_{C2} - \mu_{C1}) \tag{8B.1}$$

where k is the threshold unit cost that the healthcare provider is prepared to pay to obtain a unit increase in effectiveness. Suppose that the objective of the study is to show that Treatment 2 is more cost-effective than Treatment 1 (if this is actually true). Therefore, the null hypothesis is $H_0: \theta \leq 0$ against the alternative $H_1: \theta = \theta_R$, $(\theta_R > 0)$. The net benefits can be estimated by

$$\hat{\theta} = k(\bar{x}_{E2} - \bar{x}_{E1}) - (\bar{x}_{C2} - \bar{x}_{C1}) = (k\bar{x}_{E2} - \bar{x}_{C2}) - (k\bar{x}_{E1} - \bar{x}_{C1}) \tag{8B.2}$$

According to the central limit theorem, we can assume that, with sufficient sample sizes, the sample mean cost and efficacy under each treatment are approximately bivariate and normally distributed as follows:

$$\begin{pmatrix} \bar{x}_{Ej} \\ \bar{x}_{Cj} \end{pmatrix} \sim \text{MVN}\left[\begin{pmatrix} \mu_{Ej} \\ \mu_{Cj} \end{pmatrix}, \begin{pmatrix} \sigma_{Ej}^2/n_j & \rho_j\sigma_{Ej}\sigma_{Cj}/n_j \\ \rho_j\sigma_{Ej}\sigma_{Cj}/n_j & \sigma_{Cj}^2/n_j \end{pmatrix} \right] \tag{8B.3}$$

By considering the two treatment groups as independent, the variance of the estimated net benefits is

$$\text{Var}(\hat{\theta}) = k^2\left(\frac{\sigma_{E1}^2}{n_1} + \frac{\sigma_{E2}^2}{n_2} \right) + \left(\frac{\sigma_{C1}^2}{n_1} + \frac{\sigma_{C2}^2}{n_2} \right) - 2k\left(\frac{\rho_1\sigma_{E1}\sigma_{C1}}{n_1} + \frac{\rho_2\sigma_{E2}\sigma_{C2}}{n_2} \right) \tag{8B.4}$$

Since $\hat{\theta}$ is a linear combination of two bivariate normal variates, therefore,

$$\hat{\theta} \sim N(\theta, \text{var}(\hat{\theta}))$$

We will conclude that Treatment 2 is more cost-effective than Treatment 1 if the one-sided p-value is less than or equal to α; where $P(\text{reject } H_0 \mid \theta \leq 0) = \alpha$.

We also fix P(conclude experimental is more cost-effective than control | $\theta = \theta_R$) = $1 - \beta$. Under these two conditions, the sample size satisfies the following relationship:

$$\frac{1}{\text{Var}(\hat{\theta})} = \left(\frac{z_\alpha + z_\beta}{\theta_R}\right)^2 \tag{8B.5}$$

where $\theta_R = k(\mu_{E2R} - \mu_{E1R}) - (\mu_{C2R} - \mu_{C1R}) = k\Delta_{ER} - \Delta_{CR}$ is the hypothesised net benefits. The values of z_α and z_β are the upper 100α and 100β percentile of the standard normal distribution, respectively, where z_β corresponds to a required power of $1 - \beta$.

Substituting the variance of the net benefits from Equation 8B.1 and assuming equal sample sizes for each arm of the trial ($n_1 = n_2 = n_{CE}$) gives the sample size formula derived by Sarker & Whitehead (2004) as

$$n_{CE} = \left(z_\alpha + z_\beta\right)^2 \frac{\left[k^2\left(\sigma_{E1}^2 + \sigma_{E2}^2\right) + \left(\sigma_{C1}^2 + \sigma_{C2}^2\right) - 2k\left(\rho_1\sigma_{E1}\sigma_{C1} + \rho_2\sigma_{E2}\sigma_{C2}\right)\right]}{\left[k\Delta_{ER} - \Delta_{CR}\right]^2} \tag{8B.6}$$

They also showed that the sample sizes for unequal treatment allocation ($n_1{:}n_2 = 1{:}r$) are as follows:

$$n_1 = \left(z_\alpha + z_\beta\right)^2 \frac{\left[r\left(k^2\sigma_{E1}^2 + \sigma_{C1}^2 - 2k\rho_1\sigma_{E1}\sigma_{C1}\right) + \left(k^2\sigma_{E2}^2 + \sigma_{C2}^2 - 2k\rho_2\sigma_{E2}\sigma_{C2}\right)\right]}{r\left[k\Delta_{ER} - \Delta_{CR}\right]^2}$$

$$n_2 = \left(z_\alpha + z_\beta\right)^2 \frac{\left[r\left(k^2\sigma_{E1}^2 + \sigma_{C1}^2 - 2k\rho_1\sigma_{E1}\sigma_{C1}\right) + \left(k^2\sigma_{E2}^2 + \sigma_{C2}^2 - 2k\rho_2\sigma_{E2}\sigma_{C2}\right)\right]}{\left[k\Delta_{ER} - \Delta_{CR}\right]^2}. \tag{8B.7}$$

Definitions for the terms in equations are provided in Table 8B.1. The appropriate value of $\theta_R = k\Delta_{ER} - \Delta_{CR}$, usually referred to as the smallest clinically important difference, is much debated. Because the interventions are being compared with respect to net benefit (i.e. health outcomes net of cost), one could argue that any increase is clinically important, resulting in an arbitrary small θ_R and an arbitrary large sample size. However, in the case of costs associated with adopting experimental treatment, these would have to be offset by a sufficiently large INB, paid back over future patients. Willan (2011) shows that if C_a is the cost of adopting experimental treatment, and N is the number of future patients who would receive it if adopted, then the smallest value of INB that would offset the adoption cost is C_a/N. Therefore, setting $\theta_R = C_a/N$ and $\beta = 0.5$, he derives the sample size per group as follows:

$$n = \frac{N^2 z_\alpha^2 k^2 \left(\sigma_{E1}^2 + \sigma_{E2}^2\right) + \left(\sigma_{C1}^2 + \sigma_{C2}^2\right) - 2k\left(\rho_1 \sigma_{E1} \sigma_{C1} + \rho_2 \sigma_{E2} \sigma_{C2}\right)}{C_a^2} \tag{8B.8}$$

For the justification of assuming $\beta = 0.5$, see Willan (2011).

Technical Appendix 8C: Comparison with Briggs and Tambour's (2001) Approach

Briggs & Tambour (2001) present the following sample size formula, by considering a correlation between the cost and effect differences and assuming that $n_1 = n_2 = n_{BT}$:

$$n_{BT} = \left(z_\alpha + z_\beta\right)^2 \frac{\left[k^2\left(\sigma_{E1}^2 + \sigma_{E2}^2\right) + \left(\sigma_{C1}^2 + \sigma_{C2}^2\right) - 2k\rho\sqrt{\left(\sigma_{E1}^2 + \sigma_{E2}^2\right)\left(\sigma_{C1}^2 + \sigma_{C2}^2\right)}\right]}{\left[k\Delta_{ER} - \Delta_{CR}\right]^2} \tag{8C.1}$$

where:

$$\text{Cov}\left(\bar{x}_{E2} - \bar{x}_{E1}, \bar{x}_{C2} - \bar{x}_{C1}\right) = \frac{\rho}{n_{BT}}\sqrt{\left(\sigma_{E1}^2 + \sigma_{E2}^2\right)\left(\sigma_{C1}^2 + \sigma_{C2}^2\right)} \tag{8C.2}$$

In Equation 8B.6, Sarker & Whitehead (2004) consider the cost and effectiveness at the patient level for each treatment arm: therefore,

$$\text{Cov}\left(\bar{x}_{E2} - \bar{x}_{E1}, \bar{x}_{C2} - \bar{x}_{C1}\right) = \frac{\rho_1 \sigma_{E1} \sigma_{C1}}{n_1} + \frac{\rho_2 \sigma_{E2} \sigma_{C2}}{n_2} = \frac{1}{n_{CE}}\left(\rho_1 \sigma_{E1} \sigma_{C1} + \rho_2 \sigma_{E2} \sigma_{C2}\right)$$

For the equal correlation $\rho_1 = \rho_2 = \rho$, this result becomes

$$\text{Cov}\left(\bar{x}_{E2} - \bar{x}_{E1}, \bar{x}_{C2} - \bar{x}_{C1}\right) = \rho\left(\frac{\sigma_{E1} \sigma_{C1}}{n_1} + \frac{\sigma_{E2} \sigma_{C2}}{n_2}\right) = \frac{\rho}{n_{CE}}\left(\sigma_{E1} \sigma_{C1} + \sigma_{E2} \sigma_{C2}\right) \tag{8C.3}$$

If we consider $\sigma_{E1} = \sigma_{E2} = \sigma_E$ and $\sigma_{C1} = \sigma_{C2} = \sigma_C$, then the covariance is equal to

$$\text{Cov}\left(\bar{x}_{E2} - \bar{x}_{E1}, \bar{x}_{C2} - \bar{x}_{C1}\right) = \rho\sigma_E\sigma_C\left(\frac{1}{n_1} + \frac{1}{n_2}\right) = \frac{2\rho\sigma_E\sigma_C}{n_{CE}} = \frac{2\rho\sigma_E\sigma_C}{n_{BT}} \quad (8C.4)$$

and hence the two sample size formulae become identical in this special case as

$$n_{CE} = n_{BT} = 2\left(z_\alpha + z_\beta\right)^2 \frac{\left[k^2\sigma_E^2 + \sigma_C^2 - 2k\rho\sigma_E\sigma_C\right]}{\left[k\Delta_{ER} - \Delta_{CR}\right]^2} \quad (8C.5)$$

Therefore, when the standard deviations of the cost and efficacy are equal, Briggs & Tambour's (2001) sample size formula is a special case of this when the correlation coefficient of the two treatment arms is equal.

9

Mixed Treatment Comparisons, Evidence Synthesis

9.1 Introduction

This chapter introduces an area of current research which covers two broad areas: mixed treatment comparisons (MTCs), sometimes called network meta-analysis (NMA) and evidence synthesis, which concern pooling together data from several randomised trials (or other studies) and estimating a type of average treatment effect directly or indirectly.

9.2 MTCs

Randomised controlled trials (RCTs) that have two or more treatments are an effective yet expensive and lengthy method of comparing treatment effects. In some cases, there is simply not enough time or resource to perform additional trials to compare a new treatment with existing comparators. In such situations, an alternative method is to carry out an MTC, which uses existing data in a summarised form. The International Society of Pharmacoeconomics and Outcomes Research (ISPOR) describes an MTC as a network in which some of the pairwise comparisons have both direct and indirect evidence (Hoaglin et al., 2011). This means that some pairwise comparisons can be carried out in a given trial (direct comparisons), but using additional data can help to obtain more precise estimates. For example, in an RCT comparing A versus B and another separate trial comparing A versus C, the comparison of B versus C can be made indirectly.

There are two pieces of information for Treatment A (A vs. B and A vs. C), and a single piece of information for each of B and C (Figure 9.1). Treatment B versus C is the indirect comparison.

Pharmaceutical companies may not be able to conduct a direct comparison of a new treatment against the standard treatment or 'best' treatment

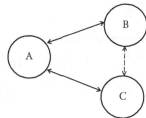

FIGURE 9.1
Diagram representing direct comparison for A vs. B and A vs. C (solid lines) and the indirect comparison for B vs. C (dashed line).

for practical or commercial reasons. This may be because either there is no standard treatment or it is commercially risky or expensive. Therefore, an MTC can be the only way to compare treatments for estimating effects which may otherwise not be possible. MTCs can be described as an extension of the models for pairwise meta-analyses. They are slightly more complex in that they can be applied to connected networks of arbitrary size and complexity. Some definitions of interrelated concepts when conducting MTCs are provided in the following subsections.

9.2.1 Direct Comparison

A direct comparison is described as a head-to-head RCT of pairs of treatments under investigation; for example Treatment A versus Treatment B, as in Figure 9.1.

9.2.2 Indirect Treatment Comparison (ITC)

An indirect comparison occurs when two or more treatments of interest are compared using a common comparator. For example, in Figure 9.1, Treatment A is the comparator in common which facilitates an indirect comparison between Treatments B and C. Indirect comparisons are also commonly referred to as adjusted or 'anchored' indirect comparisons.

9.2.3 Meta-Analysis

A meta-analysis is carried out when there is more than one trial involving the same pairwise comparisons. It combines the treatment effects (or other results) from these trials and presents one overall or pooled measure of effect size.

9.2.4 Network of Evidence

A network of evidence is a description of all the trials that include treatments of interest or trials including the comparator treatments, or both. A network diagram (Figure 9.2) gives a visual representation of all the direct comparisons that have already been made between the treatments, and can make it easier to determine the potential for indirect comparisons. It can

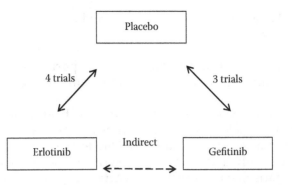

FIGURE 9.2
Network meta-analysis diagram for Table 9.1.

be as detailed or as simple as required, that is, it need not include all the treatments in the network but just those that are of interest. An NMA is a more general term for describing MTCs and indirect comparisons, and can be defined as an analysis in which the results from two or more trials that have one treatment in common are compared. Jansen et al. (2011) provide a useful summary of these.

9.2.5 Assumptions for Carrying Out MTCs

There are several assumptions to take into consideration when carrying out an MTC analysis. First, it is important to bear in mind that randomisation only holds for the original trial and not across the RCTs. Unlike meta-analyses, in which results might be pooled despite strong differences in patient characteristics and differences in the circumstances in which outcomes were collected, MTCs assume homogeneity across trials for (indirect) effects to be plausible. Hence, the homogeneity assumption must be satisfied.

The similarity assumption (Jansen, 2011) states that if trials differ among the direct comparisons (e.g. a trial comparing A vs. B differs from a trial comparing A vs. C) and these differences are due to 'modifiers' (see the following list for examples) of the relative treatment effects, then the estimate of the indirect treatment effect is biased. To explain this, consider two trials, one of which involves a comparison of Treatments A versus B (Trial 1) and the other compares A versus C (Trial 2). If Trial 1 includes patients within the age range 20–30 years and Trial 2 includes only patients within the age range 60–80 years, then age is an effect modifier. Hence, any indirect comparison will produce a result for which potential confounding can bias the indirect estimate of B versus C. Thus, to avoid violations of the homogeneity and similarity assumptions, trials must be comparable on effect modifiers such as

- Patient characteristics
- The way in which the outcomes are defined and measured

- Protocol requirements, such as disallowed concomitant medication
- Length of follow-up

In summary, with regard to NMA, it is important to check whether the trials are sufficiently similar to yield meaningful results for the ITC and MTC.

Further, the consistency assumption must also hold. This means that trials must be comparable in effect modifiers *and* there must be no discrepancy between direct and indirect estimates. Due to the way it is defined, this assumption can only, and need only, be verified when there are direct and indirect comparison data available for particular pairwise comparisons. If the consistency assumption does not hold, the indirect estimates may be biased.

In actual MTC analysis, however, since there is always sampling error, a strict evaluation of consistency based on point estimates would not be appropriate.

Lastly, an assumption of *exchangeability* should be satisfied. Exchangeability means that the relative efficacy of a treatment would be the same even if it were carried out under the conditions of any of the other trials that are included in the indirect comparison. In practice, this assumption cannot be conclusively verified, though clearly, verification of the similarity assumption would give more reason for the exchangeability assumption to hold.

All these assumptions must be considered very carefully when selecting studies for inclusion in any sort of NMA, because the validity of indirect comparisons depends on both the internal validity and the similarity of the trials conducted or reported.

Example 9.1: Verifying Assumptions for an MTC

A simple MTC was planned to compare erlotinib versus gefitinib. erlotinib versus placebo were compared in one trial (Lee, 2012), whereas gefitinib versus placebo were compared in a separate trial (Thatcher, 2005).

In order to determine whether the assumptions in Section 9.2.5 were likely to hold, attention was paid to ensuring that where information was available, the trials were comparable regarding the effect modifiers mentioned in the same section. In particular, the male to female ratio was checked; the median age was compared between the three treatment groups; and all the primary outcomes were measured in the same way: outcomes for all studies were measured by progression-free survival (PFS) and overall survival (OS). Other factors, such as whether the trials were double-blinded multicentre trials and patient recruitment per region, were also checked. However, there are often other contributing factors, such as adherence to the research protocol and quality of staff conducting the study, which might increase heterogeneity between studies.

9.3 Meta-Analysis

Meta-analysis is a statistical technique for combining the results from independent studies. On the one hand, poorly designed and conducted meta-analyses can produce biased results due to the inclusion of irrelevant studies or the exclusion of relevant studies. On the other hand, a well-conducted meta-analysis providing a transparent review of data can be more informative than simple narrative reviews.

Eligibility criteria for studies should be stated clearly, and there should be a thorough literature review. It is preferable for two independent observers to carry out the literature search and extract the data to avoid errors. Subsequently, rating the quality of the available studies based on defined criteria and blinding reviewers to the names of the authors and their institutions, the names of the journals, sources of funding and acknowledgements may produce more consistent ratings. Results from meta-analyses are often reported with effect sizes along with their confidence intervals (CIs) from individual studies, often displayed in a 'forest plot', which is a useful way of showing heterogeneity between trials.

Two modelling approaches can be taken when performing a meta-analysis (this also applies to MTCs): the fixed and random effects approaches. The fixed effect approach is based on the assumption that each of the individual studies aims to estimate the same true treatment effect (i.e. the underlying effect), and that differences between studies are solely due to chance. In other words, all studies making the same pairwise comparison should give exactly the same treatment effect, and any differences are only due to sampling error.

Another way of thinking about the fixed and random effects models is that if the true treatment effect is constant across all trials, then a fixed effects model is needed. Alternatively, if the trial-specific treatment differences are from a common distribution, one may use a random effects model. When some degree of heterogeneity is present between the studies, a random effects model might seem more appropriate for the analysis.

Example 9.2: Meta-Analysis of Seven Published Trials in Lung Cancer

A total of seven trials (Table 9.1) were identified which compared erlotinib versus placebo or gefitinib versus placebo for the PFS and OS end points. Details of the trials can be found in the references section.

A meta-analysis using a Bayesian approach was carried out for this analysis. The model assumed was of the form $\mu_i = \alpha + \beta_i$, where each hazard ratio (HR) in study i is $N(\mu_i, p_i^2)$, where p_i is the precision = $1/s_i{}^*s_i$, s_i is the standard error of each HR for each study and α_i is the overall treatment effect and β_i the study-specific effects, respectively. Effect sizes for a meta-analysis have different magnitudes of estimation error; therefore,

TABLE 9.1

Summary of Results from Seven Published Trials of Lung Cancer Treatments

Study	Author (year)	Treatment	Comparator	OS HR	PFS HR
GP1	Thatcher (2005)	Gefitinib (n = 1129)	PL (n = 563)	0.89	0.82
GP2	Zhang (2012)	Gefitinib (n = 148)	PL (n = 148)	0.84	0.42
GP3	Gaafar (2011)	Gefitinib (n = 86)	PL (n = 87)	0.81	0.61
EP1	Shepherd (2005)	Erlotinib (n = 488)	PL (n = 243)	0.70	0.61
EP2	Capuzzo (2010)	Erlotinib (n = 438)	PL (n = 451)	0.81	0.71
EP3	Herbst (2005)	Erlotinib (n = 526)	PL (n = 533)	0.99	0.94
EP4	Lee (2012)	Erlotinib (n = 332)	PL (n = 332)	0.93	0.81

Note: OS HR: hazard ratio for overall survival; PFS HR: hazard ratio for progression-free survival; PL: placebo; GP: gefitinib vs. placebo comparison; EP: erlotinib vs. placebo comparison. Studies GP1–GP3 compare gefitinib vs. placebo, while studies EP1–EP4 compare erlotinib versus placebo.

weighted means and variances are used to estimate population effect sizes. In a fixed effects model, these weights are

$$V_i = 1 / \sigma^2$$

where σ^2 is the variance of the ith effect size (for study i).

In a random effects model, the weights used are

$$W_i = 1 / (\sigma^2 + \tau^2)$$

where τ^2 is the population variance of the effect sizes. Since τ^2 is not known, it is estimated from the observed sample effect sizes (Hedges & Olkin, 1985). In this example, however, we ignore the issue of weights to avoid complexity.

9.3.1 Prior Distribution

In a Bayesian meta-analysis, the basic model assumes that α and β are distributed normally using, for example, a vague prior, such as

$$\alpha \sim N(0, 0.00001)$$

$$\beta \sim N(0, \tau)$$

where $\tau = 1/\sigma^2$.

The WINBUGS code for this random effects meta-analysis model is

```
model {
for (i in 1:N) {
t[i] <- 1 / (SE[i]*SE[i])
y[i] ~ dlnorm(mu[i], p[i])
```

```
beta[i] ~ dnorm(0, tau)
mu[i] <- alpha + beta[i]
} # end of loop
alpha ~ dnorm(0,0.00001)
tau ~ dgamma(0.001, 0.001)
}
# end of model
```

From this code, the mean (pooled) HR across the four Erlotinib versus Placebo trials was 0.8688. Similarly, for Gefitinib versus Placebo, it was 0.8664. It is noticeable that these pooled HRs are very similar, which means that, based only on the likelihood function of the current data, both treatments have approximately similar effect sizes when compared with placebo. The reader may wish to compute the estimates for the PFS end point as an exercise. In practice, the HRs are more likely to be log normally distributed rather than normally distributed. This is left as an exercise (hint: the prior distribution of τ can be considered to be a gamma).

The SAS codes for a random effects meta-analysis model using a Bayesian approach are shown in Appendix 9A.1. The data structure for a random effects meta-analysis assumes a data set with three columns: a column for the study (seven trials), a column for the log hazards ratio and a column for the standard error. The analysis would be separate for the Erlotinib versus Placebo (n = 3) and Gefitinib versus Placebo (n = 4).

Example 9.3: An MTC from Seven Published Trials in Lung Cancer

Using the data from Table 9.2, we now conduct an MTC to estimate the treatment effect for Gefitinib versus Erlotinib. Three additional studies could also have been included (one Phase II study, one retrospective multicentre study and another single-centre retrospective study, all RCTs). However, only the seven Phase III RCTs in Table 9.2 (reproduced in Sections 9.1 and 9.2 with 95% CI) are used.

A network diagram for the data in Table 9.1 is shown in Figure 9.2.

A meta-analysis was carried out in Example 9.1 to combine the pairwise comparisons and obtain one overall measure of effect size. Rather than using just the pooled estimates in the analysis, the treatment effects from each study are used in the MTC to avoid loss of information.

Step 1: Identify the measure of effect. In this case, it is the HR, and therefore we will work on the log hazards scale.

Step 2: Determine whether a Bayesian fixed effects or a random effects model is to be used. We will choose a random effects model.

Step 3: Identify the network structure. This also depends on whether pooled estimates or the individual effect sizes from each trial are

TABLE 9.2

Lung Cancer Data for Example 9.3

Study	Treatment	OS HR (95% CI)	PFS HR (95% CI)
[a]GP1	Gefitinib (n = 1129)	0.89 (0.77–1.02)	0.82 (0.73–0.92)
[b]GP2	Gefitinib (n = 148)	0.84 (0.62–1.14)	0.42 (0.33–0.55)
[c]GP3	Gefitinib (n = 86)	0.81 (0.59–1.12)	0.61 (0.45–0.83)
[d]EP1	Erlotinib (n = 488)	0.70 (0.58–0.85)	0.61 (0.51–0.74)
[e]EP2	Erlotinib (n = 438)	0.81 (0.69–0.94)	0.71 (0.62–0.82)
[f]EP3	Erlotinib (n = 526)	0.99 (0.85–1.15)	0.94 (0.85–1.03)
[g]EP4	Erlotinib (n = 332)	0.93 (0.93–1.10)	0.81 (0.68–0.95)

Note: CI: confidence interval; OS HR: hazard ratio for overall survival; PFS HR: hazard ratio for progression-free survival.

[a] Thatcher et al. (2005).
[b] Zhang et al. (2012).
[c] Ghaffar et al. (2011).
[d] Shepherd et al. (2005).
[e] Cappuzzo et al. (2010).
[f] Herbst et al. (2005).
[g] Lee et al. (2012).

used. For example, a meta-analysis earlier generated pooled effects of the direct pairwise treatment effects. These could be used, but, as noted earlier, we use the 'raw' effects (treatment differences) available from Table 9.1.

Step 4: Specify the model assumptions:

Let the log HR for a study be pooled from a normal distribution with mean μ_i and a precision that is determined by the standard errors (SE_i) of the studies (subscript i is for each study).

Let y_i be the log HR for study i, and t_i the precision of the distribution for the log HRs; then the likelihood for the model is

$$y_i \sim N(\mu_i, t_i)$$

where $t_i = 1/[SE_i * SE_i]$.

There are three treatment effects, TE(1), TE(2) and TE(3), associated with the treatments Placebo, Erlotinib and Gefitinib, respectively, each with the following identical non-informative priors:

- TE(1) ~ N(0, 0.00001)
- TE(2) ~ N(0, 0.00001)
- TE(3) ~ N(0, 0.00001)

Step 5: Identify the data to be used in the MTC.

These data include seven HRs (in Table 9.1) and the seven standard errors. It is important to ensure that the direction of the comparison is correct. For example, in Table 9.1, the HR for Gefitinib vs. Placebo in study 1 is 0.89. The HR for Placebo vs. Gefitinib would be $1/0.89 = 1.12$ and the $\log$ (HR) $= 0.11653$, which would be used in the analysis.

Step 6: Write the code (WINBUGS or SAS)

WINBUGS code:

```
# MODEL
model {
for (i in 1:N) { # indexes studies
t[i] <- 1 / SE[i]*SE[i]
y[i] ~ dnorm(mu[i], t[i]) # Likelihood function
mu[i] <- (TE[t[i]] - TE[t2[i]])
} # end i loop
# OUTPUTS
for (base in 1:(NT-1)) { # indexes treatments
for (comp in (base+1):NT) { # indexes comparators
theta[base,comp] <- exp(TE[base] - TE[comp])
}} # end base, comp loops
# PRIOR DISTRIBUTIONS
for (k in 1:NT) {
TE[k] ~ dnorm(0.0,0.00001)
} # end k loop
} # END MODEL

#DATA
list(N=7, NT=3, HR=c(0.1165338163, 0.1743533871,
0.2107210313,
0.3566749439, 0.2107210313, 0.01005033585, 0.07257069283),
SE=c(0.072, 0.155, 0.164, 0.098, 0.079, 0.077, 0.088),
t=c(1, 1, 1, 1, 1, 1, 1), t2=c(3, 3, 3, 2, 2, 2, 2))
```

Results from WINBUGS

Comparison	OS HR	PFS HR
Erlotinib vs. Placebo	0.855	0.777
Gefitinib vs. Placebo	0.874	0.419
Erlotinib vs. Gefitinib	**0.996**	**0.544**

The indirect comparison of Erlotinib versus Gefitinib results in a very slight advantage (likely to be clinically irrelevant) for Erlotinib. However, the PFS effect from Gefitinib suggests that it would offer a potentially more cost-effective treatment. One would also expect quality of life during the progression-free period to be better with Gefitinib than with Erlotinib, which would influence the cost per quality-adjusted life year (QALY).

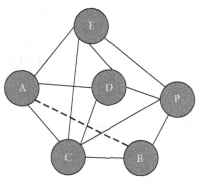

FIGURE 9.3
Studies with direct comparisons are joined by a straight line, and the indirect comparison of interest is shown with a dashed line. (Based on Scott DA, Boye KS, Timlin L, Clark JF, Best JH. *Diabetes Obes. Metab.* 15(3): 213–23, 2013.)

Example 9.4: An MTC Using SAS

In this example, an MTC is carried out using SAS code. The example is taken from Scott (2012), using data from 22 studies in type 2 diabetes. There were six treatments:

- Exenatide once weekly (ExQW) (A)
- Liraglutide 1.2 mg (B)
- Liraglutide 1.8 mg (C)
- Insulin glargine (D)
- Exenatide (bid) (E)
- Placebo (P)

The response outcome was continuous treatment effect (HbA1c measure). A useful plot was presented (Figure 9.3). The SAS data set should have four columns:

Study	Treatment	Treatment Effect	SE
1	1	−0.56	0.11
1	3	−1.44	0.17
etc.			

In Figure 9.3, the network diagram shows the direct comparisons from 22 published articles. The comparison of interest is the indirect A versus B (ExQW vs. Liraglutide), for which no direct comparison exists. The SAS code for a Bayesian MTC is shown in Appendix 9A.2, and some results are shown in Table 9.2.

9.3.2 Results (Partial Extract)

Among the six treatments, the main comparisons of interest were against placebo, and also some indirect comparisons between treatment groups. In the SAS code in Appendix 9A, the statement zero = "Placebo" monitor = (Treat) would be modified to estimate the indirect comparison of insulin

TABLE 9.3

Output from Bayesian MTC Using SAS (SAS Code in Appendix 9A.2)

Comparison versus Placebo	Mean Difference
Insulin glargine (D vs. P)	−0.819
Liraglutide 1.2 mg (B vs. P)	−1.037
ExQW versus Liraglutide 1.2 mg[a] (A vs. B)	−0.138

[a] Indirect comparison, since this comparison was not made (see Figure 9.3).

glargine versus Liraglutide. These comparisons are not present in Figure 9.3. Readers may wish to confirm the other effects using data published in Scott (2012) (Table 9.3).

Exercises for Chapter 9

1. What is the difference between an MTC and a meta-analysis, and why are these methods important for some economic evaluations?

2. What assumptions are needed for an MTC to be undertaken for an economic evaluation?

3. By adding another treatment, X, to the network in Figure 9.3, where X was compared with Treatments P and A (X vs. P and X vs. A with mean treatment effects of −0.98, SE = 0.34 and −1.63, SE = 0.65, respectively), recompute the revised estimates for the network using either WINBUGS or SAS for A versus B. Interpret your results. (Data from Scott, 2012.)

Appendix 9A: SAS/STATA Code

9A.1 SAS Code for a Bayesian Random Effects Meta-Analysis for Example 9.2

```
proc mcmc data=MTC outpost=nlout seed=276 nmc=50000 thin=5
 monitor=(HR Pooled);
 array HR[3]; **change for gefitinib vs placebo**
 parms mu tau2;
 prior mu ~ normal(0, sd=1); **state prior mean and SD**;
 prior tau2~ igamma(0.001,s=0.001); **state prior mean and SD**
 random theta ~n(mu, var=tau2) subject=studyid;
```

```
HR[studyid]=exp(theta);
Pooled=exp(mu);
model logy ~ n(theta, var=sigma2);
run;
```

9A.2 SAS Code for a Bayesian MTC for the Data in Example 9.4 Based on Scott (2012)

```
proc mcmc data=MTC outpost=nlout seed=276 nmc=50000 thin=5
 monitor=(std) outpost=outp7 dic;
 random study ~ normal (0, var=1000) subject=study init=(0);
 random treat ~normal(0, var=1000) subject=treat init=(0);
 zero ="Placebo" monitor=(Treat);**all treatments compared to
placebo**;
 *zero = Insulin glargine" monitor=(Treat);
 parms std 0.25;
 prior std ~ uniform (0,1);
 Random resid ~ normal(0, sd=std/sqrt(2)) subject = _obs_
init=(0);
 Mu=study +treat+ resid;
 model y ~ normal(mean=mu, sd=se);
 run;
```

9A.3 STATA Code for a Bayesian MTC

There are some relatively new routines found in http://cran.r-project.org/web/packages/netmeta/index.html following methods by Rücker (2012) and Krahn et al. (2013). The suite called MVMETA can be invoked from the STATA environment. Two-arm trials can invoke METAREG (standard meta-regression).

10

Cost-Effectiveness Analyses of Cancer Trials

10.1 Introduction

In this chapter, cost-effectiveness analyses using patient-level data from cancer trials will be discussed. First, the costs and their impact on evaluating cost-effectiveness are briefly reviewed. Secondly, we consider how data from cancer trials are prepared and analysed for cost-effectiveness purposes, including non-parametric and parametric survival models as well as the specific peculiarities of handling data from cancer trials. We also discuss some important statistics, such as the restricted mean and the relationship between transition probabilities and survival probabilities. We also use data from clinical trials to fit a class of models known as cubic splines (also referred to as flexible parametric models) and discuss treatment switching.

10.1.1 Costs of Cancer Care

Cancer is one of the leading causes of adult deaths, accounting for 7.6 million deaths worldwide in 2008 (Idiok et al., 2014). Lung cancer is the leading reason for cancer-related death, accounting for nearly 1.4 million deaths worldwide every year, and has an annual incidence of over 41,000 in the United Kingdom, where the total annual cost of treating lung cancer alone in 2012 was £2.4 billion. The cost of cancer care in the United States alone has been reported to be in excess of $100 billion. The annual per patient cost to the UK National Health Service (NHS) of treating cancers is £9071 for lung cancer, £2756 for bowel cancer, £1584 for prostate cancer and £1076 for breast cancer, making the average healthcare spend on each cancer patient in the United Kingdom about £2776. Cancer (and in particular lung cancer) therefore represents a significant burden of illness worldwide, and healthcare resource utilisation is likely to remain high for these patients. With such a high economic burden, it is unsurprising that the cost-effectiveness of some expensive cancer treatments is required so that decisions about scarce resources can be made in optimal ways.

Although there are several reasons for the large costs involved in cancer care, one particular cost that draws attention is the cost of treatments under investigation. These costs have not gone unnoticed in the media (Figure 10.1).

Not all cancer treatments are perceived to demonstrate benefit. Some cancer treatments are not recommended for prescribing because their 'value', when assessed through formal cost-effectiveness analyses, is not adequate. The definition of 'sufficient' is set by policymakers and can also be based on empirical assessments (e.g. by collecting all the incremental cost-effective ratios [ICERs] that have ever been reported and using the mean or median),

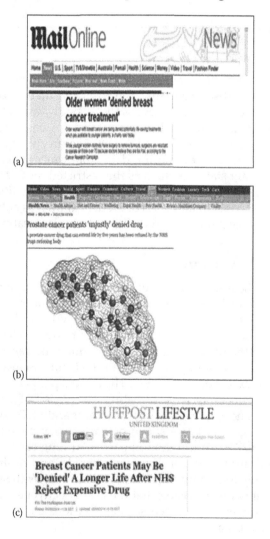

FIGURE 10.1

(a–c) Example news headlines highlighting that some cancer treatments are not being made available to patients, possibly because they are thought not to be cost-effective.

or it might be a more subjective judgement. Consequently, some cancer patients may feel that they have been 'denied' access to such treatments.

The value of treatments is also dependent on the price of the drugs (different from the cost of the drugs). The position of the ICER in the cost-effectiveness plane could move as a result of changing the price as well as improving costs and effects. In any case, this section is primarily about the more technical aspects of evaluating the cost-effectiveness of cancer treatments, and we will not dwell further on the arguments between media, payers, politicians and cancer patients.

Determining the cost-effectiveness of cancer treatments raises some particular issues:

1. In some cancers, patient survival time is short, and consequently quality-of-life (QoL) data for generating utilities are not always available. QoL after disease progression is not often collected because upon progression, patients are considered to have 'terminated' the trial. Hence, the QoL between disease progression and death is often unknown. This is particularly important because quality-adjusted life years (QALYs) are weighted over the survival time before and after disease progression. Post-progression QoL is often worse, and consequently QALYs might be overestimated.

2. When patients are randomised, it is possible that some patients may deteriorate before the start of treatment. Traditional intent to treat (ITT) analyses may include these patients in an all randomised patients analysis. However, economic evaluation is based on costs that *actually* happened or any future (or opportunity) costs. In such a situation, neither type of costs is plausible.

3. Where crossover occurs, that is, where patients randomised to the control group, upon disease progression, switch to the experimental treatment, this can over- or underestimate overall survival (OS). If the QALY gain comes from extended survival (rather than improvements in QoL), then the QALY will be unrealistic in the presence of crossover. This is discussed in Section 10.7.

4. From an economic perspective, costs and benefits over the lifetime of patients are of greatest interest. In cancer trials, the duration of follow-up is limited because not all patients have the event of interest (often death). Following up all patients until death occurs can be impractical and also very costly (e.g. staff need to be kept in employment to monitor, collect and clean additional data such as adverse events). Consequently, all patients might be censored at some convenient point (e.g. 2 years after the last patient has been recruited) so that the results of the trial can be reported. Censoring or defining a date of data cut-off does not mean an end to expected costs and effects over the entire lifetime of patients (among those

still alive) and may lead to the underestimation of costs and effects. Extrapolation methods can therefore be used to model survival costs and effects until death.

10.2 Modelling Patient-Level Data from Cancer Trials for Cost-Effectiveness

In general, for any patient-level cost-effectiveness analysis, we need to identify inputs for an economic evaluation:

1. Resource use: Used for determining costs.
2. Health-related quality of life (HRQoL): Often through generic (EQ-5D) measures.
3. Efficacy: Often OS measured from the time from first dose/randomisation until death or last known time point the patient was alive. Progression-free survival (PFS) is defined as the time from first dose until death or first progression (whichever occurs first). The definitions vary for efficacy end points, which define treatment effects from the start of randomisation. Both should be reported, because the ICER can be quite sensitive to small changes in the denominator of the ICER, particularly if the time between randomisation and start of treatment is long. If treatment administration is due to the complexity of the treatment delivery process, then both costs and effects can be considered to start from the point of randomisation.

For costs, it is important that all start and stop dates of important treatment-related adverse events are collected (because the duration of adverse events can be an important contributing cost). These can be restricted to Grade 3 and above events (instead of all adverse events). All aspects of drug compliance should be carefully monitored and reported (including wastage and reasons for wastage). Details of plausible additional medications should be recorded, including start and stop dates. This is particularly important if the patient 'crosses over' to the experimental treatment upon progression, because the date when the additional treatment starts or the date of crossover can be used as a censoring time point (experimental treatment costs may be high if switched over) in some types of analyses.

For the estimation of OS and PFS, three approaches are considered:

1. Kaplan–Meier methods (non-parametric)
2. Semi-parametric (Cox proportional hazards [PH] model)
3. Parametric models, including the flexible parametric models

For all three methods, when using patient-level data for a cost-effectiveness analysis, we require three important measures: mean survival time, median survival times and hazard ratios (HRs). The mean survival time is used for the cost-effectiveness analyses. However, this mean is a restricted (or truncated) mean, because estimates are made based on data up to the last observed survival time. The restricted mean is computed by calculating the area under the survival (Kaplan–Meier) curve. The HR is used for estimating transition probabilities between health states such as 'death' and 'progression free'. The median survival time is reported for comparison purposes (even though it is not directly used in an economic evaluation). It is instructive to understand some aspects of modelling survival data at this point. A useful reference on modelling survival data is Collett (2003).

10.2.1 Kaplan–Meier Methods

The Kaplan–Meier method does not involve modelling, and estimates of survival rates are based on the so called Nelson–Aalen estimates. The survival rates at each time point are estimated by calculating (conditional) probabilities. Hence, for a set of time points (which can be chosen), starting at time 0 (time of randomisation or start of treatment), the proportion of patients who are alive are used in the calculations. The proportion alive at each time point takes into account patients who are still alive for a given time interval.

The Kaplan–Meier estimates provide median and other quantiles for survival times. A median survival time of 6 months means that 50% of patients have died by 6 months and 50% are still alive. The restricted mean is then adjusted by the utilities obtained from HRQoL data. A separate Kaplan–Meier estimate of the mean OS and PFS is generated for each treatment group. An example SAS code for the Kaplan–Meier analysis is

```
PROC Lifetest data=<dataset>;
Time survival*censor(0); **censored=0**;
Strata treatment;
Test treatment; **log rank test**;
Run;
```

10.2.2 Semi-Parametric Methods

When it is of interest to relate survival time with covariates, a semi-parametric model called the Cox PH model can be used. The Cox model depends on the validity of a PH assumption. Violation of this assumption is often visible when survival curves cross each other or are not parallel. When this happens, it suggests that the risk of death (or the event of interest) is not constant over time. For example, in a surgery trial, there is a high risk of death initially after surgery, which then falls – this is an example of non-constant hazards (hazard = risk of death) or changing hazards. Another example might be where survival differences between treatments for males and females (e.g. females live longer) are noticeable. Therefore, adjusted measures of treatment effect can be computed.

The Cox PH model can be used to adjust for covariates when estimating the HR. The HR is an estimate of the treatment effect. An HR of 0.70 for Treatment A versus B means that patients are 30% less likely to have the event of interest (e.g. death) on Treatment A than on Treatment B. An analysis from a Cox PH model will not give estimates of median or mean survival times. Cox PH models are used because, as with Kaplan–Meier methods, one does not need to make assumptions (other than the PH assumption) about survival times.

The semi-parametric Cox PH model has continued to dominate the analysis and reporting of survival data for over 40 years. One reason is the simplicity of estimating the relationship between covariates and the hazard rate while not having to make (sometimes unjustifiable) assumptions about the baseline hazard rate.

The baseline hazard corresponds to the chance of an event (e.g. death) when all the explanatory variables assume a value of 0. The baseline hazard function can be thought of as similar to the intercept in linear regression. When the covariate assumes a value of 0 (e.g. if the control arm was given a value of 0), the baseline rate could be interpreted as the risk per unit time of death for an individual who does not take the new treatment (i.e. who takes the control).

Despite the widespread use of the Cox PH model, there are nevertheless limitations, particularly when the PH assumptions are violated. Where modelling the time dependency of a covariate (such as an individual for whom smoking status changes over time) and prediction of survival rates are necessary, the Cox PH model may have some limitations. In practice, the Cox PH model is used for estimating HRs (or adjusted HRs) and little else.

10.2.3 Parametric Methods

Where more information is required, the baseline hazard becomes important for modelling survival rates. For example, if the baseline hazard is different between hospitals, this might provide useful information on reasons why differences between treatments are observed. Prediction of survival rates beyond the last observed survival time point (extrapolation) allows health benefits (and associated costs) to be computed over the entire lifetime of patients for economic evaluation. The model fit can be greatly improved by using the baseline hazard in a statistical model. Modelling survival data can therefore be accomplished using parametric survival models, the more common of which are now discussed.

10.2.3.1 Weibull Model

The Cox PH model is often adopted when the probability distribution of the sampled survival times is unknown or it might be complicated to fit a model to the data. In the Cox PH model, the hazard function for a given patient i is

$$h_i(t) = \phi(x_i)h_0(t)$$

where:
  $h_0(t)$ is the baseline hazard function
  $\phi(x_i)$ is a function of explanatory variables x_i

For example, if the only variable is treatment, then $\phi(x_i) = \phi(x_1)$. The baseline hazard function is the hazard function in the absence of covariates. In the case of a clinical trial with a single explanatory variable (treatment) with two treatments (active and control), $h_i(t)$ is

$$h_i(t) = \exp(bx_i) \times h_0(t)$$

where:
  x_i = 1 for the active treatment
  x_i = 0 for the control

When the treatment effect is estimated, the $h_0(t)$ terms cancel out (through the partial likelihood). Where a parametric model is used, one should consider whether or not a PH parametric model is needed.

The Weibull function is a common parametric survival model employed when the PH assumption holds, and the baseline hazard function $h_0(t)$ is given by

$$h_0(t) = \lambda\gamma t^{g-1}$$

where λ and γ (both >0) are scale and shape parameters, respectively. Therefore, the hazard rate for an individual patient i consists of two parts: the baseline hazard (i.e. $h_0(t)$) and the hazard function $\phi(x_i)$. In the case of the Cox PH model, the hazard function is

$$\phi(x_i) = \exp(\Sigma\beta_i x_i)$$

Therefore, the Weibull hazard function for an individual i is

$$h_i(t) = \exp(\beta_1 X_1 + \beta_2 X_2 + \cdots \beta_k X_k) \times \lambda\gamma t^{g-1}$$

Parametric models of this type can be a useful starting point to model survival data and in many cases might be adequate. The survival function for the Weibull is

$$S(t) = \exp(-\lambda t^a)$$

where t is the survival time.

This function, given estimates for λ (here λ is the hazard rate and not the cost-effectiveness threshold) and α, will provide predictions of survival times at time t once the shape and scale parameters are estimated (when $\alpha = 1$, this becomes an exponential model). These can be estimated using the following SAS code:

```
Proc lifereg data=<data>;
Model survival*censor(0) = treatment / dist=weibull;
Run;
```

In addition, the following SAS code can be used to simulate data from a Weibull model:

```
data weibull;
do sim=1 to 1000;**1000 survival times**;
do a= 1 to 10 by 0.5;**shape parameter**;
do b= 0.1 to 10 by 0.1;**scale parameter**;
ttp = rand("WEIBULL", a, b) ;
output; output ; output;end; end; end;
run;
```

For simulating exponential survival times for given hazard rates (the ratio of two hazard rates is the HR), the following SAS code does this for each treatment group.

```
data expos1;**group 1**;
 do sim=1 to 1000;
 do x1=ranuni(213);
 simos=(-1*log(1-x1))/0.32;
 treat=1;
 output;
 end; end;
run;
data expos2;**group 2**;
 do sim=1 to 1000;
 do x2=ranuni(293);
 simos=(-1*log(1-x2))/0.62;
treat=2;
 output;
 end; end;
run;
```

In addition to the Weibull, other models such as the Gompertz model can also be used. The approach to determining which model is the most useful fit can be determined by the smallest value of the Akaike Information

TABLE 10.1

Summary of Survival Analysis Methods

Model	PH	Useful for Extrapolation	Statistic	AFT
Kaplan–Meier	No	No	Median, mean, survival rate	No
Exponential	Yes	No	HR	Yes
Weibull	Yes	Yes	HR, survival rate	Yes
Gompertz	Yes	Yes	HR, survival rate	No
Log-normal	No	Yes	HR, survival rate	Yes
Log-logistic	No	Yes	HR, survival rate	Yes
Generalised gamma	No	Yes	HR, survival rate	Yes

Criterion (AIC). In general, where the survival rates are to be estimated, an accelerated failure time (AFT) model form would be most useful. In general, the PH models the hazard rate, whereas the AFT form of the models is used for modelling the survival rates over time. Klein (2005) presents a table of hazard functions and survival functions for each model. Table 10.1 summarises the uses of each model.

10.3 Flexible Parametric Survival Models

Standard parametric models (e.g. Weibull) can be used to model survival data, but not all will retain the useful assumption of PH or even fit the observed survival pattern well. Flexible parametric modelling methodology, as expounded by Beck & Jackman (1998) and Royston & Parmar (2002), is based on using 'flexible' polynomial functions, which are fitted piecewise. This consists of several functions that are joined together at 'knots' (which act as constraints), so that the overall fitted function is smooth. The idea can be compared with linear spline fitting, but instead of linear splines (which are polynomial functions in the first degree), 'curves' are joined together to fit the observed data.

The approach to flexible parametric modelling involves joining pairs of data points using higher degrees of polynomial functions (e.g. quadratic for order two and cubic for order three, or fractional polynomials). One useful property of using splines is that the underlying functional form does not need to be known (Kruger, 2003). Splines can be used in the context of a Cox PH model to smooth the hazard function (Sleeper & Harrington, 1990); however, in this section we present the use of splines for fully parametric models. The SAS macro presented can also be used for fractional polynomials.

Although the Weibull model is a useful alternative to the Cox PH model, especially when the assumptions around distributions are reasonable, a more flexible approach (Royston & Lambert 2011) by directly modelling the baseline hazard function $h_0(t)$ as a polynomial function has been shown to be a versatile approach to fitting smooth survival functions.

10.3.1 Applications in Cancer Surveillance

Cancer registries increasingly collect important prognostic factors of survival useful for oncologists, policymakers and others to make decisions on expected survival rates for subgroups of patients. Calculating patient- and strata-specific survival rates is not easily accomplished using Cox PH modelling. For example, calculating the survival probability for an individual (or group of individuals) who is female, aged 50 (on average), with Stage III cancer is a particularly useful statistic for policymakers and planners, especially for future costing of cancer treatments.

Latymer (2013) and Bagust (2014) discuss approaches to extrapolating data in survival models, including a selection process as a guide for modelling. Several issues merit consideration.

Firstly, it is recognised that reimbursement decisions depend on having reliable evidence with respect to costs and effects accrued beyond what is often presented, because trial follow-up is curtailed for practical or submission purposes. Secondly, methods for selecting or checking models/model fit and assumptions (such as PH) should be assessed using appropriate plots and information criterion methods (see Example 10.4). Thirdly, there is debate as to whether standard parametric functions should be considered as the first choice (e.g. Weibull, log-normal), although it is agreed that the data should determine the most appropriate method through a systematic approach to modelling. In addition, a pre-specified cut-point should be identified for when the extrapolation should start. Often, this is the last observed time point or is dependent on the date of data cut-off, which itself is dictated by a regulatory constraint which requires a definition in the protocol as to when the 'end of trial' occurs. Lastly, consideration should be given to the physiological process. It is possible to fit a mathematical function to survival data which shows an excellent fit, but the estimates of the parameters may not have any physiological meaning.

10.4 Modelling Survival Data Using a Flexible Parametric Model (Example 10.1)

10.4.1 Background

The data from Lee (2012) will be used to fit a flexible parametric model using SAS. Nearly all (98%) of the patients had died at the time of analyses (658 deaths). The objective is to fit a parametric survival curve to each treatment group to predict the survival pattern and mean survival time. The results from STATA and SAS are compared. Although in this example the benefits of extrapolation are negligible, the exercise shows that a flexible parametric model can show a better fit than standard models. The Royston–Parmar (RP) models can be fitted with SAS as well as STATA.

This analysis used OS as the time to event variable; *OS_event* as the censoring variable (using a value of 1 for censored); *trt* as the treatment variable (coded as 1 or 0); and patient identifier (*patientid*). Hence, only four variables were used as inputs for the SAS macro call.

10.4.2 SAS Macro Call

Starting with a data set (named 'topic') that contains the four variables identified in Section 10.4.1, the following sequence of macros is called

1. %*sas_stset*(topic, os_event(1), os, patientid); **data set up**
2. %*sas_stpm2*(trt, scale = hazard, df = **xx**); **execution of RP-model*
3. %*predict*(surv, survival, at = trt:xx); **predicted survival rate*

The first call sets up a standard data set for survival analysis and generates a data set called '_events_'. This is followed by execution of second and third calls for the main R–P models to be implemented (%*sas_stpm2*). The variable 'trt' (for treatment group) is coded as 1 or 0, and 'df = xx' refers to the number of knots used. If df = 1, this is considered be equivalent to a Weibull model. The %predict macro estimates and saves the predicted survival rates from the previous execution. The 'trt: xx' refers to the treatment for which predicted survival rates are required ('trt:1' for Erlotinib, for example).

10.4.3 Results

The results from the SAS macro are virtually identical (Table 10.2) to the STATA output, which confirms that the SAS macro is performing as expected. The empirical survival curve (Figure 10.2) is also plotted, along with the Weibull model. The Kaplan–Meier (solid line) is approximated well by the R–P (dotted line) with three knots. The Weibull (dashed line) is a slightly worse fit.

The Weibull is a commonly employed model to compute the mean survival time (area under the survival curve) in economic evaluations for calculating QALY; in this example, the mean survival time for Erlotinib versus placebo was 6.95 versus 6.53, 6.96 versus 6.47 and 7.05 versus 6.62 months for the Kaplan–Meier, R–P(1)* and Weibull models, respectively. The Weibull model might therefore overestimate mean survival time and hence QALYs (assuming QoL is the same between treatment groups).

---

* R–P(1) refers to fitting the Royston-Parmar class of parametric models using a single knot.

TABLE 10.2

Estimates of Coefficients and Hazard Ratios from STATA and SAS

	Cox PH	Weibull[a]	RP(1)[b]
HR (SE) (STATA)	0.95 (0.08) (0.95 [0.08])	0.93 (0.09)	0.95 (0.08)
		(0.93 [0.09])	(0.95 [0.08])
Lower 95% (STATA)	0.82 (0.82)	0.78 (0.78)	0.82 (0.82)
Upper 95% (STATA)	1.11 (1.11)	1.11 (1.11)	1.11 (1.11)
Predicted 6 month		40.1 vs. 38.4	34.1 vs. 32.4
survival[c] (STATA)		(40.1 vs. 38.4)	(34.1 vs. 32.4)
AIC	7340	3818	2165

[a] Using PROC LIFEREG in SAS.
[b] RP(1): Flexible parametric using one knot.
[c] Observed 6 month survival rates for Erlotinib and Placebo were 39.1% and 36.6%, respectively.

10.5 Cost-Effectiveness of Lenalidomide (Example 10.2)

10.5.1 Background

The data for this analysis was kindly provided by Popat et al. (2014). Despite Lenalidomide and Dexamethasone being highly effective, their use in multiple myeloma (MM) is restricted due to lack of evidence of cost-effectiveness. In this example, we performed a cost-effectiveness analysis by using an alternate day dosing strategy whereby Lenalidomide commenced at 25 mg daily. Due to haematological toxicity, instead of 15 mg daily, 25 mg was given every 2 days (alternate days). Subsequent reductions of 5 mg were made for further toxicity (i.e. 15 mg, then 10 mg on alternate days instead of 10 mg, then 5 mg daily). The objective was, therefore, to compare the cost-effectiveness of Lenalidomide and Dexamethasone alternate dosing (L + $D_{Alternate}$) versus Lenalidomide plus Dexamethasone standard dosing based on the summary of product characteristics (SPC): (L + D_{SPC}).

In this example, the 'non-linear' pricing structure of Lenalidomide was exploited to provide *preliminary* cost-effectiveness evidence that giving less drug on average would result in a more cost-effective treatment without losing efficacy. Table 10.3 shows the costs of the study drug for varying dose levels.

10.5.2 Design, Patients and Treatment Schedule

This was a retrospective review of patients with relapsed MM treated with Lenalidomide and Dexamethasone in a single UK centre (University College London Hospitals Foundation Trust [UCLH]). The main efficacy end points were time to progression (TTP), PFS and OS.

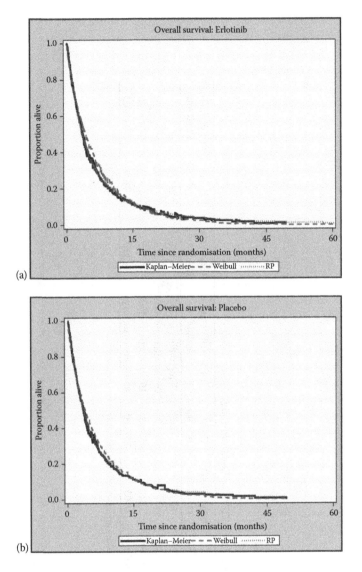

FIGURE 10.2
Comparison of predicted survival rates with (a) Erlotinib and (b) Placebo from the Weibull and flexible parametric models applied to data from the TOPICAL trial (using SAS).

10.5.3 Costs

Actual drug costs were calculated by the number of capsules prescribed over the treatment period based on the following costs per capsule (British National Formulary cost + 20% value-added tax [VAT]) as of May 2011: 25 mg = £249.60; 15 mg = £226.80; 10 mg = £216.00; 5 mg = £204.00. The cost

TABLE 10.3

Dosing Schedule and Drug Costs of Lenalidomide

Dose Level	[a]L + D$_{SPC}$	Total Dose/ Cycle (mg)	Cost/ Cycle (£)	Alternate Day Schedule	Total Dose/ Cycle (mg)	Cost/ Cycle (£)	Cost Saving/ Cycle (mg)
Starting	25 mg daily	525	5241.60	25 mg daily	525	5241.60	0
1	15 mg daily	315	4762.80	25 mg alternate days	275	2745.60	2017.20
2	10 mg daily	210	4536.00	15 mg alternate days	165	2494.80	2041.20
3	5 mg daily	105	4284.00	10 mg alternate days	110	2376.00	1908.00

[a] Dosing as described in the summary of product characteristics.

of Dexamethasone was minimal compared with that of Lenalidomide and assumed to be the same for both treatment schedules. Observed drug costs were then compared with expected costs under the SPC schedule using published data (Hoyle et al., 2008). Cost-effectiveness analysis was based on the Evidence Review Group (ERG) report for the National Institute for Health and Care Excellence (NICE) (Hoyle et al., 2008). We assumed that additional costs such as monitoring and those incurred for adverse events were the same for both schedules.

10.5.4 Cost-Effectiveness Modelling

Step 1: Patient-level data were available for a single cohort of $L + D_{Alternate}$ patients (N = 39), who were given an alternate dosing regimen. Summarised data for Lenalidomide alone, Dexamethasone alone and $L + D_{SPC}$ were taken from published data.

First, a Kaplan–Meier curve was generated for TTP for $L + D_{Alternate}$, and a Weibull parametric function was used to predict survival probabilities beyond 21 months (Figure 10.3a, reproduced below). The values of the scale and shape parameters were 0.817 and 0.0276, respectively.

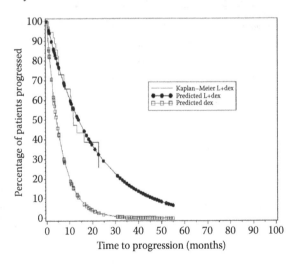

Step 2: For the comparison of $L + D_{Alternate}$ with Dexamethasone, since no patient-level data were available for Dexamethasone alone, these data had to be simulated. The mean TTP for Dexamethasone alone was simulated using published data which compared $L + D_{SPC}$ versus D using a reported HR of 0.427. It was assumed that the TTP followed an exponential model (see Table 10.4). The SAS code used to simulate from an exponential distribution with a HR of 0.427 was shown earlier. We now have a TTP survival curve for both

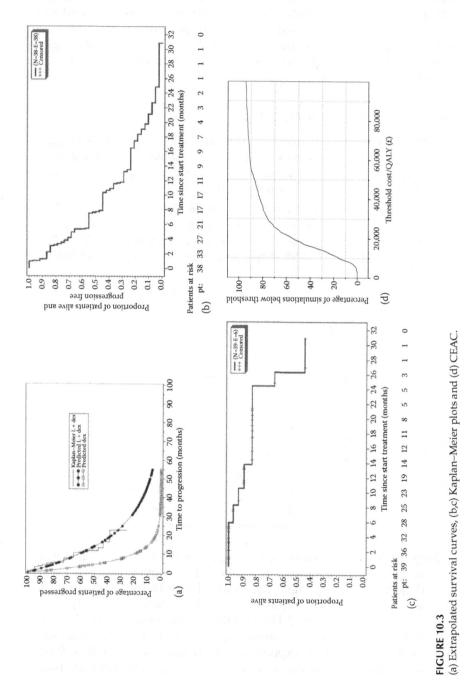

FIGURE 10.3

(a) Extrapolated survival curves, (b,c) Kaplan–Meier plots and (d) CEAC.

TABLE 10.4

Summary of Model Inputs for Health Economic Evaluation

	Parameter	$L + D_{SPC}$	Dexamethasone	$L + D_{Alternate}$
Efficacy	Median OS[1] (years)	2.8	1.6	2.8
	Mean OS[2] (years)	4.0	2.3	4.0
	Median TTP (years)	0.87[1]	0.4	0.98[3]
	Mean TTP (years)	1.7	0.6[5]	1.74
	Mean post-progression survival	2.3	1.7	2.3
Utility	Pre-progression	0.81	0.81	0.81
	Post-progression	0.64	0.64	0.64
QoL	QALY[6]	2.85	1.57	2.85
Costs	Drug	£57,921[7]	£109[7]	£29,866[8]
	Monitoring	£2,504	£404	£2,504
	Adverse events	£746	£181	£746
Total costs		£61,171	£694	£33,116
Incremental effects		$L + D_{SPC}$ vs. D		$L + D_{Alternate}$ vs. D
QALY		1.28		1.28
Costs		£60,447		£32,422
ICER (cost/QALY)		£47,247		£25,329
Base case ICER		£47,247		£25,329

[1] From ERG report (Stadtmauer et al., 2009).
[2] Modelled assuming an exponential distribution.
[3] Observed in these data.
[4] Using a Weibull function for the observed data and assumed to be the same for the standard regimen. The observed TTP was modelled using a Weibull model with a scale parameter of 0.817 and a shape parameter of 0.0276. The log likelihood test for model fit suggested that the model fit was appropriate.
[5] Estimated using TTP/log(2), assuming Dexamethasone TTP is exponential (Stadtmauer et al., 2009, p. 88).
[6] Calculated as $(1.7 \times 0.81) + (2.3 \times 0.64)$ for the L + D arm; $(0.6 \times 0.81) + (1.7 \times 0.64)$ for Dexamethasone.
[7] From ERG report, Table 27 (excluding VAT).
[8] Computed as £1,450,665.60/39 less 20% VAT based on the observed.

$L + D_{SPC}$: lenalidomide and dexamethasone standard dosing; $L + D_{Alternate}$: lenalidomide and dexamethasone alternate dosing.

$L + D_{Alternate}$ and Dexamethasone alone. The mean survival time is computed for each of these. The next step was to estimate the OS for both groups.

Step 3: Using the observed OS data for $L + D_{Alternate}$, assuming an exponential model to fit the observed Kaplan–Meier curve, we can extrapolate beyond the last time point to the maximum time an 'average' human being could theoretically live – set to 100 years. Consequently, the mean OS for $L + D_{Alternate}$ was 4 years.

In order to estimate the mean OS for Dexamethasone alone, we used the fact that OS = TTP/log(2) (also used by the ERG). For the

Dexamethasone group, therefore, the mean TTP was estimated to be 2.3 years. Similar data for L + D_{SPC} were already reported (Table 10.4).

Step 4: Using reported data for the pre- and post-progression utilities and the fact that OS = pre-progression survival + post-progression survival, the QALY can be estimated (Table 10.4); 10,000 simulations were carried and the mean (of the simulated means) was used for each model input, as shown in Table 10.4.

Step 5: After computation of the mean effects, the mean costs were taken from published data and computed for the experimental group. All future costs and effects were discounted at a rate of 3.5% per year and excluded VAT. Estimates of utility from generic measures of QoL were obtained from published data (none were collected in the cohort of patients on alternate dosing, but utilities were assumed to be the same, even though there was an expectancy that with reduced dosing, the toxicity burden would be improved).

Many of the data were provided, and these are simple arithmetic computations. The base case ICER reported for reimbursement (in the original health technology assessment [HTA] submission) was £47,247. For the alternate day dosing, this was £25,329/QALY. Clearly, there is substantial value in an alternative dosing strategy. However, it rests on the assumption that mean survival times will be similar – which would best be investigated in a randomised controlled trial (RCT). However, recall that this analysis offers *preliminary* evidence of cost-effectiveness of an alternative dosing schedule. The cost-effectiveness acceptability curve (CEAC) and other plots are shown in Figure 10.2.

The ICER for the alternate day dosing regimen estimated in this analysis is well below that of the estimate provided by the ERG report (the ERG is an external independent group that confirms the original cost-effectiveness analysis submitted by the pharmaceutical company) of >£47,000 for the L + D_{SPC} regimen. Probabilistic sensitivity analysis for uncertainty of model inputs resulted in a mean ICER of £31,110 (range £3,715 to >£100,000) based on 10,000 simulations for the alternate dosing. The estimated probability of cost-effectiveness of the alternate schedule at a NICE threshold of £30,000 is about 76% (Figure 10.2). Additional sensitivity analysis by varying model inputs by 20% (drug costs), 10% (TTP) and 5% (utility) resulted in ICERs varying between £23,904 and £30,047, still below or close to the £30,000 threshold used by NICE (Table 10.4).

Since this was a retrospective analysis, there are some limitations. For cost-effectiveness purposes, it was assumed that the efficacy of the two regimens was similar. This was justified because the estimate of the median OS of 2.2 years was comparable with published data of approximately 2.8 years and is likely to be an underestimate due to a limited number of death events. The median TTP in this data was estimated at 0.98 years (11.8 months) compared with 10.3 months in the published literature. Hence, estimates of efficacy were comparable, given the limitations of the study.

10.6 Transition Probabilities and Survival Rates

In this section, the relationship between transition probabilities and survival rates is briefly explored. This is important, because in some situations a Markov model is used for an economic evaluation of a cancer treatment. In some cases, the primary end points for cancer trials are survival rates at fixed time points (e.g. 2 year survival rates compared between experimental and control treatments). Therefore, change in survival rates from one time point to another has a relationship with patients transitioning (since, as fewer patients are alive over time, it is plausible some would have moved into a progressive disease state before death).

In Section 10.2.3, we saw how an exponential model had a survival function of the form $S(t) = e^{-\lambda t}$. In this expression, λ is the hazard rate or the rate of an event (such as death) per unit of time: for example five deaths per month (the unit of time). The probability can be expressed as a rate. If p is the proportion alive at time t, then $p = e^{-\lambda t}$ and hence $\lambda = -(\log(p)/t)$. For example, if after 2 years, 50 patients out of 200 are alive, and assuming the death rate is constant (the same number of deaths each year), the death rate is $[-(\log(1-0.25))]/5 = -\log[0.75]/5 = -0.287/5 = 0.0574$, and the probability of death at year 3 would be $1 -\exp(-0.0574 \times 3) = 0.158$ (15.8%).

The Kaplan–Meier method estimates probabilities of survival within given discrete intervals. If the survival rate at 2 years $= S(2) = 0.8$, then the survival rate for the year before (year 1) can be written as $S(1) = 0.9$. In general, the survival during the previous year u will be $S(t - u)$. Therefore, the current survival rate at time t is $S(t)$. We are one step closer to deriving the transition probability, because we now have an idea that the survival rates change from the previous time $(t - u)$ to the current time (t). If the survival rate is $S(t)$, then the failure (death) rate is $1 - S(t)$. The probability of surviving during an interval of time is $S(t)/S(t - u)$; hence, the probability of death (failure) during an interval of time is $1 - [S(t)/S(t - u)]$.

The transition probability for death is defined as $TRp = 1 - [S(t)/S(t - u)]$.

In survival analysis, there is a key relationship between the survivor function, the hazard function and the probability density function (PDF) (see Collett, 2003 for further details)

$$f(t) = h(t) \times S(t)$$

and further,

$$S(t) = \exp\{-H(t)\}$$

where:

    $f(t)$ is the PDF
    $S(t)$ is the survival function
    $H(t)$ is the cumulative hazard function

Therefore, the transition probability TR_p is

$$1 - \left[S(t)/S(t-u)\right] = 1 - [\exp\{-H(t)\}/\exp\{-H(t-u)\}$$

$$= 1 - \exp\{H(t-u) - H(t)\}$$

For example, if the survival times follow an exponential distribution, the PDF is

$$f(t) = \lambda \exp\{-\lambda t\}$$

where:

    $h(t)$ $= \lambda$
    $S(t)$ $= \exp\{\lambda t\}$
    $H(t)$ $= \lambda t$

Using $TR_p = 1 - [S(t)/S(t-u)] = 1 - [\exp\{-H(t)\}/\exp\{-H(t-u)\}] = 1 - [\exp\{H(t-u) - H(t)\}$,

$$TR_p = 1 - \exp\{\lambda(t-u) - \lambda t\} = 1 - \exp\{-\lambda u\}$$

This is the transition probability of moving from one time point to another. Typically, what is required in cancer trials is the transition probability of moving between health states such as death, progression and being alive over time.

Figure 10.4 shows a diagram for a typical Markov model which models transition probabilities between health states commonly seen in cancer trials (alive, death and progressive disease). One would normally generate a survival curve for PFS and OS and then generate the transition probabilities between health states. For example, if OS and PFS are both considered exponentially distributed, then a continuous time Markov process can be modelled, or alternatively, a patient-level simulation approach can be used (i.e. simulate PFS, then simulate OS ensuring PFS is <OS, and also simulate the censoring distribution). Published SAS code by Hui-min et al. (2004) is available for modelling a multi-state Markov process using a Weibull function – which can be a very complex analysis.

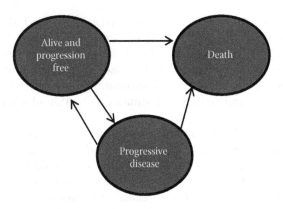

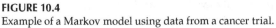

FIGURE 10.4
Example of a Markov model using data from a cancer trial.

10.7 Handling Crossover (Treatment Switching) in Cancer Trials

In Section 10.1.1, we introduced the issue of crossover in cancer trials, which can make differences between treatment arms appear larger (or smaller) and therefore bias the estimate of the QALY. Jonsson et al. (2014) and Latimer (2014) provide an overview of the issues and implications of crossover for economic evaluation. Crossover allows patients to experience the treatment they did not get because the current treatment a patient has received did not yield the desired effect (efficacy or safety). Crossover often occurs at the point of unblinding (at the patient level) and results in a loss of information. In a (double-blind) trial, at the point of progression (switching over), it would not be known whether an experimental treatment is efficacious or not – this would usually be known after the data analysis is complete or partially complete (through interim analysis).

The underlying issue with regard to crossover is how it is handled in the statistical analysis. Designing a trial for potential crossover may be impractical or very challenging (i.e. it is not always known whether patients will switch over or who they are, and it is unlikely patients would want to be randomised again to another 'choice' when they have just stopped taking the existing treatment). Moreover, some patients can be treated with experimental treatments (switch over treatment) by moving to a location where the treatment is available. For example, stereotactic whole brain radiotherapy (SBRT) offered to lung cancer patients is only available at some particular hospital sites. If some patients decided they wished to receive this despite being randomised to the alternative treatment group, this would have similar effects as crossing over, biasing treatment effects.

The statistical issues around treatment switching also have similar implications for situations in which patients take other concomitant medications, stop taking treatment or take cocktails of treatments, and also dose adjustments (which are commonly captured through dose modifications in the case report forms). The presence of significant crossover can have serious implications for QALY estimates due to biased efficacy estimates. Hence, this problem is not just a health economic evaluation question, but a more general question on generating unbiased estimates of treatment effect in the presence of crossover in cancer trials.

Therefore, in general, a definition of a crossover treatment should be identified and stated upfront. One should also determine whether crossover is likely to be a concern for a given reimbursement authority, especially for long-term treatment effects in the presence of crossover. One should identify and distinguish between spontaneous and treatment-related crossover (e.g. by patient decision, rather than due to progression). It is always better to design a trial, where possible, to avoid crossover, and perform trial design work taking into account the presence and absence of crossover. It would also be useful to have an idea of the rate of crossover expected, based on experience with previous trials.

There are several approaches to handling crossover suggested in the published literature:

(i) ITT analysis

(ii) Per protocol (PP) analysis

(iii) Including crossover treatment as a time-varying covariate in a Cox model

(iv) Inverse-probability-of-censoring weighting (IPCW)

(v) Rank-preserving structural nested failure time models (RPSFTMs)

We will discuss (iii) and (iv) in more detail, since these appear to be the currently recommended approaches (Jonsson, 2014; Morden, 2011; Ishak, 2011). No one particular method has been identified as the 'best' approach suitable for all situations. Interested readers may consult the references for specific details.

In the case of ITT, since patients are analysed according to the group they were randomised to, effects measured after switching are considered as belonging to the treatment group from which they are switching. For example, a patient is randomised to the control group (C). The patient's OS time (until progression) was 6 months, but the PFS time was 4 months. Patients in the experimental arm (E) have an average (median) OS time of 8 months and PFS time of 6 months. At 4 months, this patient switched to the experimental treatment. It appears the patient lived for 7 months.

10.7.1 ITT

Using an ITT analysis, the patient would have an OS of 7 months attributable to C. If the median OS is 8 months on E, the patient has an OS of 1 month

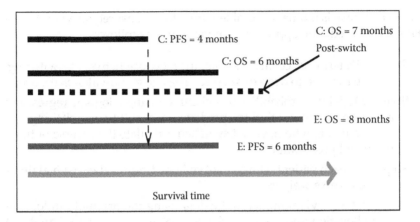

FIGURE 10.5
Relationship between switching and survival. C, censored; E, event.

less than average (Figure 10.5). The fact that the extended survival might be due to post-switch experimental treatment is not factored in. Consequently, the treatment effect is underestimated, as will be the QALY. Conversely, if a switch from E to C happened (because E was too toxic), the ITT analysis may lead to the wrong interpretation. Note that due to randomisation, the effects are not necessarily biased (in the statistical sense), because data are treated as belonging to a randomised experiment.

10.7.2 Per Protocol (PP) Analysis

A PP analysis would censor patients from the date they switched. This analysis can be subject to selection biases: patients who switch may well be those with poorer prognosis. In non-inferiority trials (not common in cancer), the PP is the primary population for analysis. In addition, a PP analysis may result in a substantially reduced sample size and consequent loss of power.

10.7.3 IPCW

The IPCW approach (Robins & Finklestein, 2000) involves calculating the probability of censoring in relation to a set of confounders (e.g. age, performance status, stage) using a logit model. We assume there are no hidden or unmeasured confounders – a difficult assumption to verify. For a given data set, we identify the date of switching for each patient. If a date is present, a switch is considered to have occurred. The logit model is then regressed against a set of covariates

$$\text{Response} = \text{intercept} + \text{covariates} \left(\text{including time-varying covariates}\right)$$

where response is a binary variable with 1 for a censored survival time and 0 if not censored. The approach to modelling is as follows:

Step 1: Determine covariates that may be predictive of switching. Decide the rule to drop or keep covariate (e.g. if p-value $\gg .05$, drop).

Step 2: Model the probability of switching (using a logistic regression) and for each patient find the predicted probability of switching. For those patients who have not switched, calculate the inverse of these predicted probabilities.

Step 3: For each patient who has not switched, adjust the survival times by the inverse weights.

Step 4: Fit a survival model to the data using the original randomisation allocation and censor patients who switch (as in a PP analysis) on the date of switching.

Step 5: Compute mean survival and QALY.

10.7.4 RPFSTM

This method uses the AFT (see Table 10.1) form of a survival model, with the objective of presenting a measure of treatment effect by adjusting for those who cross over. The approach of the RPFSTM is to consider the total survival time T_i, consisting of two parts (using Jonsson, 2014 notation): T_i before crossover on the *control arm*, referred to as T_i^{on}, and survival time after switching T_i^{off}.

Observed survival time:

$$T_i = T_i^{on} + T_i^{off} \tag{10.1}$$

In addition, we have something called the 'unobserved' survival time, U_i – the survival time that *would have been observed* on the control arm after progression if the patient had not switched – plus a measure of effect ψ (called the acceleration factor, which is considered to be a casual factor; values of $\psi < 0$ suggest a beneficial treatment effect).

Unobserved survival time:

$$U_i = T_i^{off} \exp(\psi) \times T_i^{on} \tag{10.2}$$

If $\exp(\psi) = 0.7$, for example, this is like saying that 1 year of taking the study treatment (control arm) is equivalent to 0.7 years of not taking the treatment. The value of ψ is iterated until a value is found which yields the highest p-value in a statistical test (e.g. log rank). The process of estimation is carried out through G-estimation (Robins, 2005). Assumptions for RPFSTM include assuming a common treatment effect, that is, the treatment effect from progression onwards (until death) in the control group is similar to the treatment effect from randomisation until death in the experimental group.

In addition, U_i is considered to be independent of the randomised treatment group, and there is an explicit assumption that switchers and non-switchers are comparable (or need to be comparable). Further details of this method can be found in the bibliography.

One last consideration for treatment switching relates to how PSA is conducted in the presence of switching. Where switching is of concern, the PSA and ICERs should be presented by taking into account the methods presented in Sections 10.7.1 through 10.7.4. Estimates of ICERs and PSA analysis from both the standard PSA and those as a result of adjusting for switching should be reported. The issue also extends to the extrapolation of survival curves. For example, if using IPCW, one can use the weighted survival times with a parametric survival function. Kaplan–Meier curves, too, should be presented with and without taking switching into account. This is a subject of current research.

10.8 Landmark Analysis and Presenting Survival Data by Tumour Response

The landmark responder model represents an alternative to the standard partitioned model described in Section 10.6 when PFS and OS data are immature or not available. In a landmark responder approach analysis, PFS and OS are presented for each response category as described in Figure 10.6 at a *specified time point*. The landmark (responder) model therefore consists of two adjoining mathematical models forming (i) a decision tree structure (for computing the number of responders to treatment) at the landmark time point; and (ii) a traditional three-state partitioned survival model to estimate mean PFS and OS conditional on the response status of the cohort. One important issue about landmark response models in cancer is that 'response' at a given time point is different from the commonly protocol-defined 'best response', which is based on RECIST (Response Evaluation Criteria in Solid Tumours) criteria. It is therefore important that when such landmark analyses are being carried out (often for short-term market access reasons), there is an understanding that the definition of a landmark 'visit' response and 'best response' will be different (the best response is often a confirmed tumour response at least 4 weeks apart).

In Figure 10.6, patients who respond (completely or partially) to treatment have better PFS and OS than those who have stable disease. Patients who progress at the landmark analysis point can only experience death following progression, and therefore, a PFS curve is not necessary.

Tumour response is divided into response subcategories (called health or states), as long as each category is a predictor of survival, costs or health utilities. Instead of two states (response vs. no response), the decision tree is

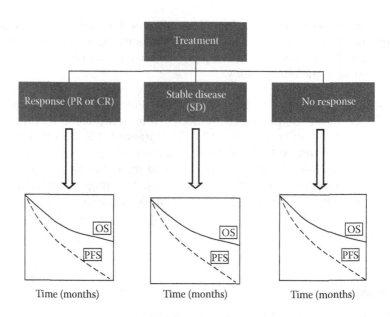

FIGURE 10.6
Responder model structure, landmark point at 3 months (12 weeks) related to the survival model.

structured into three or more response states (e.g. complete/partial response, stable disease and progression).

Patients enter the model in the 'stable disease' state. Response is measured at the landmark point, after which all patients are included in a partitioned survival analysis for economic evaluation by responder status. The starting time for the analysis of PFS and OS is the time of the landmark point, not the time of randomisation (as in a standard survival analysis).

For the landmark responder model, PFS and OS are estimated from the point at which patients respond to treatment (Figure 10.7). The choice of the landmark time point is not a simple matter, and requires input from several experts. One advantage of this approach is that tumour response is not impacted by treatment switching (crossover) because data are collected prior to the potential use of switch therapy (because they are responding, and hence, there is no need to switch). It is also easier to synthesise this type of data to build more evidence about treatment benefit. Differentiation of treatments in terms of resource use and disease management, as well as utilities, can be made within the responder part of the model by incorporating health states, such as stable disease (SD), complete response (CR) and partial response (PR), which can continue into the partitioned survival model.

The difficulty associated with this approach is that landmark-type analyses are not common in HTA submissions, and response is not often the primary end point – although there have been some treatments recently that

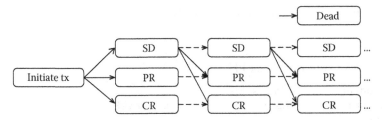

FIGURE 10.7
Responder model structure in terms of tumor response.

have been approved based on Phase II data with response as the primary end point. Moreover, response data on comparators may be too limited to conduct detailed comparative economic evaluations.

Exercises for Chapter 10

1. What are the important characteristics of economic evaluations in cancer trials?

2. What are the more commonly cited methods for extrapolating future survival in cancer trials? What are the limitations?

3. When might the use of a flexible parametric survival analysis be useful?

4. What is a landmark analysis, and why would this be used for an economic evaluation? Describe the difficulties associated with conducting an economic evaluation based on landmark analysis.

5. What are the disadvantages of undertaking an economic evaluation within each of the response (RECIST) categories (see Section 10.8)?

6. Why is crossover (treatment switching) in cancer trials problematic, and what impact can this have on the ICER?

7. Which methods could be used when undertaking an economic evaluation and crossover is present?

Appendix 10A: SAS/STATA Code

10A.1 SAS Code for a Survival Analysis

```
PROC Lifetest data=<dataset>;
Time survival*censor(0); **censored=0**;
Strata treatment;
```

```
Test treatment; **log rank test**;
Run;
```

10A.2 SAS Code for a Cox PH Model

```
PROC phreg data=<dataset>;
Model survival*censor(0)= treatment / risklimits;
**censored=0**;
Run;
```

In STATA this would be (here treatment is an indicator):

```
.stcox i.treatment
Failure _d: censor
Analysis survival _t: time1(##here _t is the survival time, _d
is the censoring variable##)
Id:id
```

10A.3 SAS Code for Fitting a Weibull Model

```
PROC lifereg data=<dataset>;
Model survival*censor(0)= treatment / dist=Weibull ;
**censored=0**;
Run;
```

10A.4 SAS Code for Simulating Data from Weibull and Exponential Distributions

```
data weibull;
do sim=1 to 1000;**1000 survival times**;
do a= 1 to 10 by 0.5;**shape parameter**;
do b= 0.1 to 10 by 0.1;**scale parameter**;
  ttp = rand("WEIBULL", a, b) ;
output; output ; output;end; end; end;
run;
```

For simulating exponential survival times for given hazard rates (the ratio of two hazard rates is the hazard ratio), the following SAS code does this for each treatment group.

```
data expos1;**group 1**;
   do sim=1 to 1000;
   do x1=ranuni(213);
      simos=(-1*log(1-x1))/0.32;
   treat=1;
      output;
   end;  end;
run;
```

```
data expos2;**group 2**;
   do sim=1 to 1000;
   do x2=ranuni(293);
      simos=(-1*log(1-x2))/0.62;
   treat=2;
      output;
   end;  end;
run;
```

10A.5 SAS Code for Performing a Flexible Parametric Model of the Royston–Parmar Class

This code is too long for this appendix and can be requested from the author: https://www.researchgate.net/profile/Iftekhar_Khan5.

10A.6 STATA Code for Royston–Parmar Model

In this example, two treatments were compared (Treatment A and B): B was coded as 0 and A as 1. The outcome for analysis was PFS and the censoring variable was pfscens. The commands sts, by(trt) and stcox trt were used to determine treatment effects with a standard survival and Cox PH model. A flexible parametric model is then chosen using 3 knot with the stpm2 command. The KM plots are fitted comparing A and B.

```
gen byte trt = arm=="A"
lab def arm 0 "B" 1 "A"
lab val trt arm

// Stset PFS data, ignoring 38 failures at time "0" (.031)
gen byte event = (pfscens == 0)
stset pfs event, enter(time .031)
sts, by(trt)
stcox trt

// Clearly no difference between treatments. Fit Flexible
Parametric Model with // scale hazard and look for suitable d.f.

forvalues j = 1 / 8 {
     qui stpm2, df(`j') scale(hazard)
     di `j', e(AIC), e(BIC)
}
// Selected model minimise both AIC and BIC, in this example.
// Best model has 3 d.f. Compare survival curve with
Kaplan-Meier
// for each treatment group.

stpm2 trt, df(3) scale(hazard)
predict s0, survival at(trt 0)
```

```
predict s1, survival at(trt 1)
sts gen S0 = s if trt==0
sts gen S1 = s if trt==1

line s0 s1 S0 S1 _t, sort sav(survival, replace)
xtitle("Months from randomisation") ///

 ytitle("Progression-free survival probability") lp(l l - -)
ysca(r(0 1)) yla(0(.25)1) ///

 leg(label(1 "Flex parametric, A") label(2 "Flex parametric,
B") label(3 "Kaplan-Meier, A") label(4 "Kaplan-Meier, B"))
```

11

The Reimbursement Environment

11.1 Regulatory Requirements for Clinical Efficacy versus Payer Requirements for Value

Previously, when a drug received marketing authorisation, the process of achieving a price for the new drug was negotiated with government bodies without any need to demonstrate the value of the new treatment. In some countries, this process was replaced by a revised paradigm (Figure 11.1) in which a payer evidence review body – that is, a group of experts – assesses whether the new treatment is cost-effective. Hence, two sets of experts, one that assesses efficacy and another that separately assesses value, review the evidence for efficacy and cost-effectiveness/reimbursement. Reimbursement authorities are essentially representatives of health systems from different countries who determine the level of compensation for the drugs that are made available for their patient populations through their respective health systems.

Figure 11.1 shows the regulatory and payer evidence process which is broadly in practice today. On the left of the dashed line, it represents the process for review of efficacy and safety data using randomised controlled trials (RCTs) (for market authorisation).

On the right-hand side of Figure 11.1, reimbursement will rest on several additional factors rather than just evidence from the single RCT submitted. Considerations of a country's practices, the choice of comparator and using additional evidence from local smaller studies will influence how different payer demands should be met. Statistical support is often disjointed between the two processes, and each statistician (one addressing efficacy and the other post-marketing needs) may have different objectives. For example, if a group sequential design is planned, emphasis on stopping rules and efficacy data to ensure a quality interim analysis is one consideration; the statistician supporting health technology assessment (HTA) submission is likely to be concerned with other forms of data that will demonstrate value more efficiently, such as ensuring longer-term follow-up, rather than stopping early – which compromises the collection of longer-term quality-of-life (QoL), utility and safety data.

Current regulatory and reimbursement process

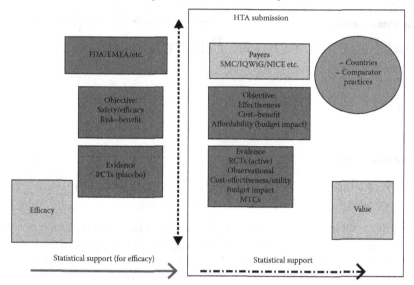

FIGURE 11.1

Previous/current reimbursement or review process.

Figure 11.2 offers a different paradigm, in which there is much greater alignment between the needs of a trial for reimbursement and demonstrating clinical efficacy.

Situations can arise (and have arisen) in which the efficacy requirements are met for all EU member states, but some or all may not pay for the new drug. This is like saying '...we appreciate the new drug is efficacious, but for *this country*, the level of *effectiveness* is not strong enough'. Consequently, some have considered that there may be a need for a centralised European pricing and reimbursement strategy (14th ISPOR Congress, 2011).

It may appear strange that a UK patient cannot get access to a new drug in the United Kingdom while it is available in three other EU member state countries. In this situation, it is likely that different judgements about value are used to determine the effectiveness of the new treatment (e.g. different comparators are considered less or more important, or the effectiveness of the new drug may also depend on additional forms of healthcare – e.g. palliative care, screening strategies). Moreover, there may be different cost-effectiveness thresholds, differences in patient mix, different drug prices and potentially 27 different estimates of relative effectiveness (even if we only need one). There are also likely to be significant political and practical challenges (e.g. the desire in some countries for less EU integration) before the idea of minimum guaranteed healthcare for every EU citizen can be offered. However, there is at least some commonality in that RCTs are to be used to demonstrate efficacy as a

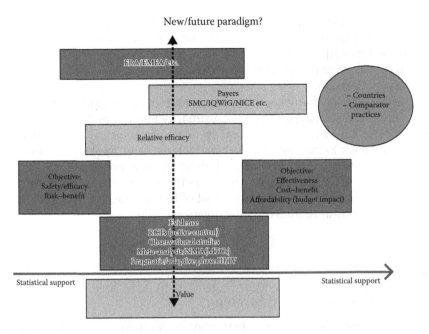

New/future paradigm?

FDA/EMEA/etc.

Payers
SMC/IQWiG/NICE etc.

– Countries
– Comparator
practices

Relative efficacy

Objective:
Safety/efficacy
Risk–benefit

Objective:
Effectiveness
Cost–benefit
Affordability (budget impact)

Evidence
RCTs (active-control)
Observational studies
Meta-analysis/NMA(MTCs)
Pragmatic/adaptive phase III/IV

Statistical support   Statistical support

Value

FIGURE 11.2
Alternative paradigm for aligning payer requirements with market authorisation requirements.

starting point (although anyone who has been involved in an EU submission process will note that interpretations of efficacy results will rarely be unanimous between country representatives).

11.2 Reimbursement and Payer Evidence Requirements across Different Countries

Given that alignment is unlikely to come any time soon, one challenge is to meet the reimbursement needs of different countries. Targeting the key market players is usually the best strategy – for example if France, the United Kingdom, Germany and Italy represented 70% of the market share, it would make sense to design a trial that meets the requirements of all these countries (e.g. recruiting patients from each country and understanding the population, baseline characteristics and any unmet needs in each country). Not all countries accept quality-adjusted life years (QALYs). QALYs are a UK requirement, a 'good to have' item in some Scandinavian countries and not necessary in countries such as France, Spain, Germany and Italy – but that is not to say that such evidence will not be considered. Therefore, it is not 'all about QALYs or the EQ-5D', as is often thought.

Table 11.1 shows a summary of the reimbursement requirements for some key EU member states. Although only one or two countries request QALYs or an incremental cost-effectiveness ratio (ICER) as a specific requirement, nearly all would find cost-effectiveness evidence useful or helpful for a reimbursement decision.

One concern for many countries is affordability or budget impact, which has not been discussed in this book, primarily because budget impact analysis focuses on factors not necessarily related to the clinical trial data for analysis – such as size and clinical characteristics of the target population (all other alternative treatments excluding the new one being considered) and costs of current treatments – to assess the affordability of the new treatment. Consequently, the cumulative cost of treating the target population is of greater importance than cost-effectiveness evidence.

11.3 Market Access and Strategy

One area which not all researchers may be familiar with is *market access* (MA). MA is a key commercial function which attempts to define a strategy to achieve a market share. Increasingly, MA departments need to work with health economists and researchers to provide marketing intelligence and a strategy for achieving the clinical trial's cost-effectiveness (and broader commercial) objectives. For example, it may not be sufficient to develop a new drug with excellent design capabilities (e.g. a capsule that can combine pain relief and deal with constipation symptoms), because a cheap opioid taken with a cheap laxative can achieve a similar effect. The MA function offers strategic vision and advice to avoid such situations. If reliable estimates of utility are not available, a separate utility study might be commissioned, or the clinical trial should ensure that these concerns are addressed. The definition of health states might also be considered before the trial starts. If effect sizes for health states are modest or small (even if clinically relevant), they may not be large enough to differentiate the new treatment from existing treatments (a clinical trial designed as a superiority trial might achieve this, whereas a non-inferiority trial may not).

11.4 Value-Based Pricing

Another area of recent interest is value-based pricing (VBP). Since vast amounts of a health budget are spent on new drugs (about 10% in the United Kingdom alone), one way of controlling pharmaceutical profits is to provide

TABLE 11.1

Cost-Effectiveness Requirements from Various Reimbursement Systems in Europe (as of 2014)

Country	Cost-Effectiveness Evidence Required and Additional Notes
Austria[a]	Pharmacoeconomic data mandatory Price given only if >50% of EU states market the drug Price of new treatment compared with comparator prices as basis of reimbursement is key
Belgium[b]	Cost–utility and cost–benefit perspective including total costs of drug for the healthcare system and patient population Reimbursement category: for example cancer is category A, analgesics or comfort drugs are category D Special budgets for cancer
Denmark[c]	Pharmacoeconomic data optional Price of new treatment compared with comparator prices as basis of reimbursement is key, and budget impact (affordability) Greater calls for use of pharmacoeconomic data as part of evidence without being specific
France[d]	Not compulsory Any data submitted are reanalysed by experts Decisions made primarily on efficacy and not cost-effectiveness evidence Budget cuts may make such data necessary in the future
Germany[e]	Emphasis on clinical benefit Pharmacoeconomic data might be used to establish maximum price Longer-term RCT data important Cost–benefit assessment needed in the form of 'efficiency frontier analysis' and budget impact analysis
Ireland[f]	Irish HTA guidelines should be followed Usually ICERs < €45,000/QALY recommended, hence cost–utility analysis Other factors also taken into account (not a black and white criterion) Positive ICER no guarantee of approval
Spain[g]	No formal requirement, but focus on reimbursement is usually based on budget impact, although cost-effectiveness analyses not ignored Autonomous regions have established local committees to assess value of drugs
Italy[h]	Submission of pharmacoeconomics not obligatory Importance of pharmacoeconomic studies being noted HTA section opened to optimise prescribing practices NICE restricted to assessing clinical effectiveness; responsibility for assessing cost-effectiveness and funding decisions to a consortia of general practitioners (GPs)
Netherlands[i]	Pharmacoeconomic studies required Cost-effectiveness from a societal perspective (including burden of illness) important Guidelines in place for cost-effectiveness analyses Reanalyses conducted, particularly for expensive drugs CVZ[i] manual with guidelines similar to those of NICE for conducting cost-effectiveness analyses

(Continued)

TABLE 11.1 (Continued)

Cost-Effectiveness Requirements from Various Reimbursement Systems in Europe (as of 2014)

Country	Cost-Effectiveness Evidence Required and Additional Notes
Poland[j]	Request for cost-effectiveness analysis, but no specific details provided Budget impact requested and considered key
United Kingdom[k]	NICE guidelines available ICERs and QALYs requested Reviews for single and multiple technology appraisals (STA and MTA) One of the leaders in HTA
Norway[l]	PE studies and cost-effectiveness mandatory for reimbursement HTA assessments reside as a subgroup within NoMA – specific guidelines available Clinical societal budget impact AEs and PROs cited as important
Sweden[m]	Pharmacoeconomic guidelines in place Health economic evaluation essential Cost-effectiveness from a societal perspective is included (e.g. loss of production, time away from work)

Note: PRO, patient reported outcomes.

Reimbursement bodies:

[a] Haupterverband der osterreichischen Sozialverischerungstrager (HVB)
[b] Comission de prix des Specialites Pharmaceutiques (CPSP)
[c] Medicintilskudsnaevnet
[d] Comission de la Transparence (CT)
[e] Institute for Quality and Efficiency in Health Care (IQWiG)
[f] National Centre for Pharmacoeconomics (NCPE) or HIQA
[g] Subdireccion general de Calidad de Medicamentos y Productos Sanitarios (SGCMPS)
[h] Pricing and Reimbursement Committee (CPR) and Italian General Medicine Society (IGMS)
[i] College voor Zorgverzekeringen (CVZ)
[j] Komisja Ekonoiczna (Economic Committee)
[k] National Institute for Health and Clinical Excellence (NICE) and Scottish Medicines Consortium (SMC)
[l] Norwegian Medical Agency (NoMA) and the blaresept ordningen (reimbursement system)
[m] Tanvards –och lakemedelsformansverket (TLV)

incentives based on the value that the new treatment offers. Hence, VBP attempts to estimate the value of a new treatment based on all the current evidence, and reimbursement is dependent on the quantification of such value. The difference from the current National Institute for Health and Care Excellence (NICE) approach is that affordability of a new treatment is based on the ICER, whereas VBP introduces more factors, such as unmet needs, innovation and how the new drug reduces the burden of the illness on society. In Sweden, for example, VBP includes equity considerations (i.e. some patients are not denied treatment), the severity of the disease and cost-effectiveness. This book has focused on the last of these (which is also the attitude of some reimbursement authorities). The limited budget perspective in the United Kingdom is that value might be based on achieving a cost/QALY <£30,000, whereas in Sweden a broad societal perspective is based on €90,000/QALY (approx. £72,000/QALY).

As methodologies for VBP advance, it will no doubt have implications for clinical development programmes and trial design – in particular for orphan drugs (since the target population is small). VBP is likely to have implications for methodology too, in particular developing thresholds and weightings for CE threshold which take into account the severity of the disease. More specifically, VBP is likely to result in some kind of cost per QALY threshold at the National Health Service (NHS) level, which would then vary according to levels of the factors mentioned in the previous paragraph (innovation, burden of illness, other societal benefits).

11.5 Submissions for Payer Evidence

In this section, we briefly describe the evidence needed for submission of evidence to NICE for a single technology appraisal. The full details on submitting evidence can be obtained from the NICE website. The purpose here is to list the key contents of Section B, Part 6: 'Cost-Effectiveness'. The submission preparation should be planned alongside the clinical development plan and protocol; there is always a chance that the new treatment will not demonstrate efficacy, and hence, there should be careful consideration of how much pre-planned work needs to be done.

After discussing the background and clinical evidence, the cost-effectiveness section should include in the submission the details in the following list. The level of detail in the content of the submission obviously depends on the complexity of the trial and the cost-effectiveness model:

- Identify all studies with respect to costs and effects. This should ideally be carried out well before the main (RCT) trial analysis is reported so that data synthesis can be accomplished in time for the final submission.
- Describe completely the treatment pathway.
- Describe the model structure, including all assumptions and justification for the model as well as its limitations. Identify all health states, describing reasons why the selected health states are suitable. Does the economic model reflect the treatment pathway?
- Describe in the model:
  - Time horizon
  - Cycle length (if appropriate, and any half-cycle corrections)
  - How health effects were measured (e.g. QALYs)
  - Discounting rate (usual 3.5% per year)
  - Perspective (NHS, societal)

- State whether interventions and comparators are in agreement with the marketing authorisation licence. Different countries might have different practices – for example some cancer therapies are given for six cycles as standard practice in some countries and for four cycles in others. This would be reflected in different costs.
- State clinical end points, both primary and secondary, definitions, frequency of measurement and justification for choice.
- State approaches to modelling end points – for example if time to progression was modelled using parametric survival models – and the reason for the selection of the model of choice (i.e. any goodness of fit tests such as Akaike's Information Criterion [AIC]).
- State measures of effect (hazard ratios, mean differences, etc.).
- Give details of handling adverse events (AEs), and types and duration of the key AEs.
- Describe how transition probabilities were calculated, stating explicit formulae: for example, if the transition probability $tp(t_{ij}) = 1-S(t + u)/S(t)$ denotes the probability of a transition occurring in the interval t to t + u and the parameters for a Weibull (for example) are $exp(-exp(-u/\sigma))*t^{1/\sigma}$, where u = intercept and σ = scale. If covariates are included, t in this formula is divided by the exponent of the sum product of the vector of coefficients and the vector of covariate values.
- State whether transition probabilities vary over time (time-dependent hazards assumed).
- Identify any surrogates.
- Give source for utilities and a description of QoL, if it is likely to change over time.
- State whether mapping was used. Describe the model and justification for its use and any sensitivity analyses.
- List other sources/studies from which utilities were obtained in the cost-effectiveness analysis (state the mean and 95% limits).
- Describe details of baseline QoL and utilities.
- Identify resources, including intervention and comparator costs, with sources for costs and unit prices.
- Describe all sources of potential uncertainty in sensitivity analyses.
- Provide results, including cost-effectiveness acceptability curves (CEACs); scatter plots; base case ICERs; summary statistics of costs and effects; analyses, including subgroup analyses and details of how the analysis was validated; details of budget impact analysis.

11.6 Further Areas for Research

Cost-effectiveness in clinical trials still offers many challenges for researchers in health economic evaluation. Some of the methods described in this book for health economic evaluation are not always understood by researchers and payer review groups. For example, the methods described for handling crossover in cancer trials can be highly complex, and can take time for even an experienced statistician to appreciate. The use of Bayesian methods in clinical trials still poses problems, since many statisticians are used to designing clinical trials using classical (frequentist) methods and are accustomed to using SAS, in which Bayesian computation has only relatively recently become slightly more feasible.

Statistical support for payer evidence is often disjointed from the main RCT; statistical support should start from Phase I and beyond. Moreover, statistical knowledge and experience of payer evidence requirements are invested in a few statisticians and other clinical trial staff, whereas the concept of demonstrating the value of treatments is more of a 'mindset' about adding *value* and not just an isolated (statistical) function. Often, payer evidence considerations are not thought about until the start of the Phase III trial, when there is a need to think about what combination of possible studies will result in a high probability of MA, or what combination of studies is likely to demonstrate value. The other challenge is that the statistical tools for eliciting value are unclear or unknown to statisticians – G-estimation techniques for handling crossover and value of information (VOI) methods are areas in which training and courses have more recently become available for application to clinical trials. Value analysis/reimbursement analysis/payer analysis plans are needed in addition to statistical analysis plans which clearly identify those end points necessary for demonstrating value and those entered as direct inputs into a health economic analysis. Figures 11.3 and 11.4 give some examples of challenges in trial design and analysis for the future.

Several other technical issues remain as interesting areas of research which require additional work, such as sample sizes based on VOI, VBP models, cost-effectiveness based on more than one end point, extrapolation in the presence of switching treatment (in cancer trials) and also simulation in the presence of crossover for probabilistic sensitivity analysis. Applications of flexible parametric models are still limited, and improvements are needed in mapping functions for health-related quality of life (HRQoL), particularly some generic measures for which the ability of generic instruments to detect meaningful treatment effects is of concern. In summary, there is no shortage of subject areas for statisticians and researchers in health economic evaluation to develop, whether in academia or industry.

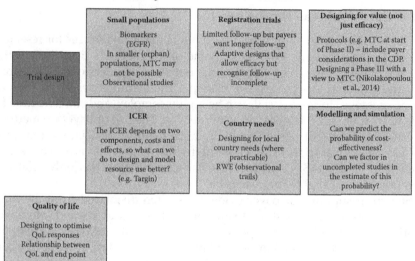

FIGURE 11.3
Some challenges in designing trials for health economic evaluation.

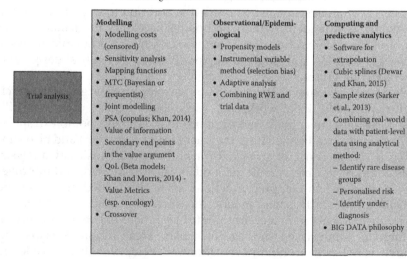

FIGURE 11.4
Some challenges in analysing data for health economic evaluation.

Exercises for Chapter 11

1. Describe the reimbursement environment in Europe. Why is it important to understand this when planning an economic evaluation?

2. What are the challenges of planning for reimbursement across multiple countries? How would this impact the way we might plan an economic evaluation?

3. What evidence is needed in an economic evaluation for a NICE submission?

4. What is value-based pricing? Why is it important?

References

Ades AE, Lu G, Claxton K. Expected value of sample information calculations in medical decision modeling. *Med. Decis. Mak.* 2004; 24: 207.

Ades AE, Madan J, Welton NJ. Indirect and mixed treatment comparisons in arthritis research. *Rheumatology (Oxford)* 2011; 50 Suppl. 4: iv5–9.

Al MJ, Van Hout BA, Michel BC, Tutten FF. Sample size calculation in economic evaluations. *Health Econ.* 1998; 7(4): 327–35.

Arnold D, Girling A, Stevens A, Lilford R. Comparison of direct and indirect methods of estimating health state utilities for resource allocation: Review and empirical analysis. *BMJ* 2009; 339: b2688.

Backhouse ME, Gnanasakthy A, Schulman KA, Akehurst R, Glick H. The development of standard economic datasets for use in the economic evaluation of medicines. *Drug Info. J.* 2000; 34(4): 1273–91.

Bagust A, Beale S. Survival analysis and extrapolation modeling of time-to-event clinical trial data for economic evaluation: An alternative approach. *Med. Decis. Making* 2014; 34(3): 343–51.

Bang H, Tsiatis AA. Estimating medical costs with censored data. *Biometrika* 2000; 87(2): 329–43.

Barton P, Bryan S, Robinson S. Modelling in the economic evaluation of health care: Selecting the appropriate approach. *J. Health Serv. Res. Policy* 2004; 9(2): 110–18.

Basu A, Manning WG. Estimating lifetime or episode-of-illness costs under censoring. *Health Econ.* 2010; 19(9): 1010–28.

Beck N, Jackman S. Beyond linearity by default: Generalized additive models. *Am. J. Polit. Sci.* 1998; 42: 596–627.

Blazeby JM, Avery K, Sprangers M, Pikhart H, Fayers P, Donovan J. Health-related quality of life measurement in randomized clinical trials in surgical oncology. *J. Clin. Oncol.* 2006; 24(19): 3178–86.

Blommestein HM, Franken MG, Uyl-de Groot CA. A practical guide for using registry data to inform decisions about the cost effectiveness of new cancer drugs: Lessons learned from the PHAROS registry. *Pharmacoeconomics* 2015. [Epub ahead of print]

Box GEP, Draper N. *Response Surfaces, Mixtures, and Ridge Analyses*, 2nd edn [of *Empirical Model-Building and Response Surfaces*, 1987]. Hoboken: John Wiley, 2007.

Bozzani FM, Alavia Y, Jofre-Bonet M, Kuper H. A comparison of the sensitivity of EQ-5D, SF-6D and TTO utility values to changes in vision and perceived visual function in patients with primary open-angle glaucoma. *BMC Ophthalmol.* 2012; 12(1): 43.

Bradbury PA, Tu D, Seymour L, Isogai PK, Zhu L, Ng R, Mittmann N, Tsao MS, Evans WK, Shepherd FA, Leighl NB. Economic analysis: Randomized placebo-controlled clinical trial of erlotinib in advanced non-small cell lung cancer. *J. Natl Cancer Inst.* 2010; 102: 298–306.

Brazier J, Rowen D. NICE DSU Technical Support Document 11: Alternatives to EQ-5D for Generating Health State Utilities; Report by the Decision Support Unit, March 2011.

Brazier JE, Yang Y, Tsuchiya A, Rowen DL. A review of studies mapping (or cross walking) non-preference based measures of health to generic preference-based measures. *Eur. J. Health Econ.* 2010; 11(2): 215–25.

Briggs A, Gray AM. Power and sample size calculations for stochastic cost-effectiveness analysis. *Med. Decis. Making* 1998; 18(Suppl.): S81–92.

Briggs A, Tambour M. The design and analysis of stochastic cost-effectiveness studies for the evaluation of health care interventions. *Drug Inform. J.* 2001; 35: 1455–68.

Briggs A, Claxton K, Sculpher M. *Decision Modelling for Health Economic Evaluation.* Oxford: Oxford University Press, 2006.

Briggs AH. *Statistical Methods for Cost-Effectiveness Research: A Guide to Current Issues And Future Developments.* London: Office of Health Economics, 2003.

Briggs AH, O'Brien BJ. The death of cost-minimization analysis? *Health Econ.* 2001; 10(2): 179–84.

Briggs AH, Wonderling DE, Mooney CZ. Pulling cost-effectiveness analysis up by its bootstraps: A non-parametric approach to confidence interval estimation. *Health Econ.* 1997; 6(4): 327–40.

British National Formulary (BNF). 2011: http://www.bnf.org/bnf/index.htm.

British National Formulary (BNF). 2012: http://www.bnf.org/bnf/index.htm.

Calman KC. Quality of life in cancer patients: An hypothesis. *J. Med. Ethics* 1984; 10(3): 124–7.

Cappuzzo F, Ciuleanu T, Stelmakh L, Cicenas S, Szczésna A, Juhász E, Esteban E, et al.; SATURN investigators. Erlotinib as maintenance treatment in advanced non-small-cell lung cancer: A multicentre, randomised, placebo-controlled phase 3 study. *Lancet Oncol.* 2010; 11(6): 521–9.

Carayanni VJ, Tsati EG, Spyropoulou GC, Antonopoulou FN, Ioannovich JD. Comparing oil based ointment versus standard practice for the treatment of moderate burns in Greece: A trial based cost effectiveness evaluation. *BMC Complement. Altern. Med.* 2011; 11: 122.

Carpenter J, Kenward M. *Multiple Imputation and Its Application.* Chichester: John Wiley, 2013.

Cassell DL. *BootstrapMania!: Re-Sampling the SAS Way.* Corvallis, OR: Design Pathways, Paper 268, 2010.

Chow SC, JP Liu. *Design & Analysis of Bioavailability and Bioequivalence Studies*, 2nd edn. New York: Marcel Dekker Publications, 2000.

Claxton K. The irrelevance of inference: A decision-making approach to the stochastic evaluation of health care technologies. *J. Health Econ.* 1999; 18(3): 341–64.

Cohen J. *Statistical Power Analysis for the Behavioral Sciences*, 2nd edn. Hillsdale: Lawrence Erlbaum Associates, 1988.

Collet D. *Modelling Survival Data in Medical Research*, 2nd edn. Survival Analysis 2003. London: Chapman & Hall.

Crott R, Briggs A. Mapping the QLQ-C30 quality of life cancer questionnaire to EQ-5D patient preferences. *Eur. J. Health Econ.* 2010; 11(4): 427–34.

Culyer AJ. *Handbook of Health Economics*, Vols IA,B and II. Amsterdam: Elsevier, 2011.

Dakin H, Wordsworth S. Cost-minimisation analysis versus cost-effectiveness analysis, revisited. *Health Econ.* 2013; 22(1): 22–34.

Deb P, Holmes AM. Estimates of use and costs of behavioural health care: A comparison of standard and finite mixture models. *Health Econ.* 2000; 9(6): 475–89.

Dolan P. Modeling valuations for EuroQol health states. *Med. Care* 1997; 35(11): 1095–1108.

Dewar R, Khan I. A new SAS Macro for flexible parametric survival modelling: Applications to clinical trials and surveillance data. *Comput Methods Prog Biomed* 2015 (*in press*).

Drummond M, O'Brien B. Clinical importance, statistical significance and the assessment of economic and quality-of-life outcomes. *Health Econ.* 1993; 2(3): 205–12.

Drummond MF, Sculpher Mark J, Torrance GW, O'Brien BJ, Stoddart GL. *Methods for the Economic Evaluation of Health Care Programmes*, 3rd edn. Oxford: Oxford University Press, 2005.

Duan N. Smearing estimate: A nonparametric retransformation method. *J. Am. Stat. Assoc.* 1983; 78: 605–10.

Dukhovny D, Lorch SA, Schmidt B, Doyle LW, Kok JH, Roberts RS, Kamholz KL, Wang N, Mao W, Zupancic JA. Economic evaluation of caffeine for apnea of prematurity; Caffeine for Apnea of Prematurity Trial Group. *Pediatrics* 2011; 127(1): e146–55.

Dunlop W, Uhl R, Khan I, Taylor A, Barton G. Quality of life benefits and cost impact of prolonged release oxycodone/naloxone versus prolonged release oxycodone in patients with moderate-to-severe non-malignant pain and opioid-induced constipation: A UK cost-utility analysis. *J. Med. Econ.* 2012; 15(3): 564–75.

Dunlop W, Iqbal I, Khan I, Ouwens M, Heron L. Cost-effectiveness of modified-release prednisone in the treatment of moderate to severe rheumatoid arthritis with morning stiffness based on directly elicited public preference values. *Clinicoecon. Outcomes Res.* 2013; 5: 555–64.

Efficace F, Bottomley A, Osoba D, Gotay C, Flechtner H, D'haese S, Zurlo A. Beyond the development of health-related quality-of-life (HRQOL) measures: A checklist for evaluating HRQOL outcomes in cancer clinical trials – Does HRQOL evaluation in prostate cancer research inform clinical decision making? *J. Clin. Oncol.* 2003; 21(18): 3502–11.

European Medicines Agency. Guideline on the investigation of bioequivalence; CPMP/EWP/QWP/1401/98, 2010.

Fairclough LD. *Design and Analysis of Quality of Life Studies in Clinical Trials*, 2nd edn. London: Chapman & Hall/CRC, 2010.

Fan X, Felsovalyi A, Sivi SA, Keenan SC. *SAS for Montecarlo Studies: A Guide for Quantitative Researchers*. Cary, NC: SAS Publications, 2002.

Fleishman AI. A method for simulating non-normal distributions. *Psychometrika* 1978; 43: 521–31.

Forbes Magazine 2014. http://www.forbes.com/sites/danmunro/2014/02/02/annual-u-s-healthcare-spending-hits-3-8-trillion/.

Gaafar RM, Surmont VF, Scagliotti GV, Van Klaveren RJ, Papamichael D, Welch JJ, Hasan B, Torri V, van Meerbeeck JP; EORTC Lung Cancer Group and the Italian Lung Cancer Project. A double-blind, randomised, placebo-controlled phase III intergroup study of gefitinib in patients with advanced NSCLC, non-progressing after first line platinum-based chemotherapy (EORTC 08021/ILCP 01/03). *Eur. J. Cancer* 2011; 47: 2331–40.

Gelber RD, Goldhirsch A. Comparison of adjuvant therapies using quality-of-life considerations. *Int. J. Technol. Assess. Health Care* 1989; 5(3): 401–13.

Gheorghe A, Roberts TE, Ives JC, Fletcher BR, Calvert M. Centre selection for clinical trials and the generalisability of results: A mixed methods study. *PLoS One* 2013; 8(2): e56560.

Glick HA. Sample size and power for cost-effectiveness analysis (part 1). *Economic Evaluation in Clinical Trials*, 2nd edn. 2011; 29(3): 189–98.

Glick HA, Doshi JA, Sonnad SS, Polsky D. Oxford: Oxford University Press, 2014.

Goodwin PJ. Reversible ovarian ablation or chemotherapy: Are we ready for quality of life to guide adjuvant treatment decisions in breast cancer? *J. Clin. Oncol.* 2003; 21(24): 4474–5.

Gray AM, Clarke PM, Wolstenholme JL, Wordsworth S. *Applied Methods of Cost-Effectiveness in Healthcare.* Oxford: Oxford University Press, 2011.

Gregorio GV, Dans LF, Cordero CP, Panelo CA. Zinc supplementation reduced cost and duration of acute diarrhea in children. *J. Clin. Epidemiol.* 2007; 60(6): 560–6.

Grieve R, Nixon R, Thompson SG, Normand C. Using multilevel models for assessing the variability of multinational resource use and cost data. *Health Econ.* 2005; 14(2): 185–96.

Grieve R, Nixon R, Thompson SG, Cairns J. Multilevel models for estimating incremental net benefits in multinational studies. *Health Econ.* 2007; 16(8): 815–26.

Grootendorst PV. A comparison of alternative models of prescription drug utilization. *Health Econ.* 1995; 4(3): 183–98.

Harper R, Brazier JE, Waterhouse JC, Walters SJ, Jones NM, Howard P. Comparison of outcome measures for patients with chronic obstructive pulmonary disease (COPD) in an outpatient setting. *Thorax* 1997; 52(10): 879–87.

Haycox A, Drummond M, Walley T. Pharmacoeconomics: Integrating economic evaluation into clinical trials. *Br. J. Clin. Pharmacol.* 1997; 43(6): 559–62.

Hedges LV, Olkin I. *Statistical Method for Meta Analysis.* New York: Academic Press, 1985.

Herbst RS, Prager D, Hermann R, Fehrenbacher L, Johnson BE, Sandler A, Kris MG, et al. TRIBUTE Investigator Group. TRIBUTE: A phase III trial of erlotinib hydrochloride (OSI-774) combined with carboplatin and paclitaxel chemotherapy in advanced non-small-cell lung cancer. *J. Clin. Oncol.* 2005; 23(25): 5892–9.

Hoaglin DC, Hawkins N, Jansen JP, Scott DA, Itzler R, Cappelleri JC, Boersma C, et al. Conducting indirect-treatment-comparison and network-meta-analysis studies: Report of the ISPOR Task Force on Indirect Treatment Comparisons Good Research Practices: part 2. *Value Health* 2011; 14(4): 429–37.

Hornberger J, Eghtesady P. The cost-benefit of a randomized trial to a health care organization. *Control. Clin. Trials.* 1998; 19(2): 198–211. Review.

Huang Y. Cost analysis with censored data. *Med. Care* 2009; 47(7 Suppl. 1): S115–19.

Hughes DA, Bagust A, Haycox A, Walley T. The impact of non-compliance on the cost-effectiveness of pharmaceuticals: A review of the literature. *Health Econ.* 2001; 10(7): 601–15.

Hui-Min W, Ming-Fang Y, Chen TH. SAS macro program for non-homogeneous Markov process in modeling multi-state disease progression. Computer methods. *Programs Biomed.* 2004; 75(2): 95–105.

Idiok-Akpan PA, Nonye AA. Perceptions of burden of caregiving by informal caregivers of cancer patients attending University of Calabar Teaching Hospital, Calabar, Nigeria. *Pan Afr. Med. J.* 2014; 18: 159 (using GLOBOCAN, Cancer Fact Sheet-Cancer Incidence and Mortality Worldwide. International Agency for Cancer Research, 2008. http://www.iarc.fr/en/media-centre/iarcnews/2010/globocan2008.php).

Institut für Qualität und Wirtschaftlichkeit im Gesundheitswesen (IQWiG) (Institute for Quality and Efficiency in Health Care). General methods; Version 3.0 of 27 May 2008.

Ishak KJ, Caro JJ, Drayson MT, Dimopoulos M, Weber D, Augustson B, Child JA, et al. Adjusting for patient crossover in clinical trials using external data: A case study of lenalidomide for advanced multiple myeloma. *Value Health* 2011; 14(5): 672–78.

Jansen JP, Fleurence R, Devine B, Itzler R, Barrett A, Hawkins N, Lee K, Boersma C, Annemans L, Cappelleri JC. Interpreting indirect treatment comparisons and network meta-analysis for health-care decision making: Report of the ISPOR Task Force on Indirect Treatment Comparisons Good Research Practices: Part 1. *Value Health* 2011; 14(4): 417–28.

Keus F, de Jonge T, Gooszen HG, Buskens E, van Laarhoven CJ. Cost-minimization analysis in a blind randomized trial on small-incision versus laparoscopic cholecystectomy from a societal perspective: Sick leave outweighs efforts in hospital savings. *Trials* 2009; 10: 80.

Kharroubi SA, O'Hagan A, Brazier JE. A comparison of United States and United Kingdom EQ-5D health states valuations using a nonparametric Bayesian method. *Stat. Med.* 2010; 29(15): 1622–34.

Khan I. Probabilistic sensitivity analyses for clinical trials of cost-effectiveness using the method of copulas: A comparison of simulation methods. SCT Meeting Abstracts. *Society for Clinical Trials.* 2014.

Khan I. Interpreting small effects from the EORTC-QLQC-30 2015. *Health Qual. Life Outcomes (in press).*

Khan I. SCET-VA® Version 1 released May 2015; Sample size software for cost-effectiveness and value of information analysis. Available from author on request at https://www.researchgate.net/profile/Iftekhar_Khan5.

Khan I, Morris S. A non-linear beta-binomial regression model for mapping EORTC QLQ- C30 to the EQ-5D-3L in lung cancer patients: A comparison with existing approaches. *Health Qual. Life Outcomes* 2014; 12: 163.

Khan I, Morris S, Hackshaw A, Lee SM. Cost effectiveness of first-line erlotinib in patients with advanced non-small-cell lung cancer unsuitable for chemotherapy. *BMJ [Open]* 2015; 5(7): e006733.

Khan I, Pashayan N, Matata B, Morris S, Maguire J. Sensitivity and responsiveness of EQ-5D-5L compared with EQ-5D-3L and EORTC-QLQC-30 in non-small cell lung cancer patients. *Health Qual. Life Outcomes (in press).*

Kobelt G. *Health Economics. An Introduction to Economic Evaluation*, 2nd edn. London: Office of Health Economics, 1996.

Kruger CJC. Constrained cubic spline interpolation for chemical engineering purposes, 2003. Available at: http://www.korf.co.uk/spline.pdf.

Lambert PC, Billingham LJ, Cooper NJ, Sutton AJ, Abrams KR. Estimating the cost-effectiveness of an intervention in a clinical trial when partial cost information is available: A Bayesian approach. *Health Econ.* 2008; 17(1): 67–81.

Laska EM, Meisner M, Siegel C. Power and sample size in cost-effectiveness analysis. *Med. Decis. Making* 1999; 19(3): 339–43.

Leal J, Gray AM, Clarke PM. Development of life-expectancy tables for people with type 2 diabetes. *Eur. Heart J.* 2009; 30(7): 834–9.

Lee SM, Khan I, Upadhyay S, Lewanski C, Falk S, Skailes G, Marshall E, et al. First-line erlotinib in patients with advanced non-small-cell lung cancer unsuitable for chemotherapy (TOPICAL): A double-blind, placebo-controlled, phase 3 trial. *Lancet Oncol.* 2012; 13(11): 1161–70.

Lewis G, Peake M, Aultman R, Gyldmark M, Morlotti L, Creeden J, de la Orden M. Cost-effectiveness of erlotinib versus docetaxel for second-line treatment of advanced non-small-cell lung cancer in the United Kingdom. *J. Int. Med. Res.* 2010; 38(1): 9–21.

Lin DY. Proportional means regression for censored medical costs. *Biometrics* 2000; 56(3): 775–8.

Lin DY, Feuer EJ, Etzioni R, Wax Y. Estimating medical costs from incomplete follow-up data. *Biometrics* 1997; 53(2): 419–34.

Liu L, Strawderman RL, Cowen ME, Shih YC. A flexible two-part random effects model for correlated medical costs. *J. Health Econ.* 2010; 29(1): 110–23.

Machin D, Campbell MJ, Tan SB, Tan SH. *Sample Size Tables for Clinical Studies*, 3rd edn. Oxford: John Wiley, 2009.

Maguire J, Khan I, McMenemin R, O'Rourke N, McNee S, Kelly V, Peedell C, Snee M. SOCCAR: A randomised phase II trial comparing sequential versus concurrent chemotherapy and radical hypofractionated radiotherapy in patients with inoperable stage III non-small cell lung cancer and good performance status. *Eur J Cancer* 2014; 50(17): 2939–49.

Maiwenn J, Van Hout B. A Bayesian approach to economic analyses of clinical trials: The case of stenting versus balloon angioplasty. *Health Econ.* 2000; 9(7): 599–609.

Manca A, Rice N, Sculpher MJ, Briggs AH. Assessing generalisability by location in trial-based cost-effectiveness analysis: The use of multilevel models. *Health Econ.* 2005; 14(5): 471–85. Erratum in *Health Econ.* 2005; 14(5): 486.

Manca A, Lambert PC, Sculpher M, Rice N. Cost-effectiveness analysis using data from multinational trials: The use of bivariate hierarchical modeling. *Med. Decis. Making* 2007; 27(4): 471–90.

Mann R, Brazier J, Tsuchiya A. A comparison of patient and general population weightings of EQ-5D dimensions. *Health Econ.* 2009; 18(3): 363–72.

Manning WG, Mullahy J. Estimating log models: To transform or not to transform? *J. Health Econ.* 2001; 20(4): 461–94.

McKenzie L, van der Pol M. Mapping the EORTC QLQ C-30 onto the EQ-5D instrument: The potential to estimate QALYs without generic preference data. *Value Health* 2009; 12(1): 167–71.

Mihaylova B, Briggs A, O'Hagan A, Thompson SG. Review of statistical methods for analysing healthcare resources and costs. *Health Econ.* 2011; 20(8): 897–916.

Morden JP, Lambert PC, Latimer NR, Abrams KR, Wailoo AJ. Assessing methods for dealing with treatment switching in randomised controlled trials: A simulation study. *BMC Med. Res. Methodol.* 2011; 11: 4.

Nikolakopoulou A, Mavridis D, Salanti G. Using conditional power of network meta-analysis (NMA) to inform the design of future clinical trials. *Biometrical J.* 2014; 56(6): 973–90.

NICE Guide to the methods of technology appraisal 2013; published 4 April 2013. http://publications.nice.org.uk/pmg9.

Nixon RM, Thompson SG. Parametric modelling of cost data in medical studies. *Stat. Med.* 2004; 23: 1311–31.

Noble S, Hollingworth W, Tilling K. Missing data in trial-based cost-effectiveness analysis: The current state of play. *Health Econ.* 2012; 21: 187–200.

O'Connor RD, O'Donnell JC, Pinto LA, Wiener DJ, Legorreta AP. Two-year retrospective economic evaluation of three dual-controller therapies used in the treatment of asthma. *Chest* 2002; 121(4): 1028–35.

O'Hagan A, Stevens JW. Bayesian assessment of sample size for clinical trials of cost-effectiveness. *Med. Decis. Making* 2001; 21: 219–30.

O'Hagan A, Stevens JW. Bayesian methods for design and analysis of cost-effectiveness trials in the evaluation of health care technologies. *Stat. Meth. Med. Res.* 2002; 11: 469–90.

Osoba D, Rodrigues G, Myles J, Zee B, Pater J. Interpreting the significance of changes in health-related quality-of-life scores. *J. Clin. Oncol.* 1998; 16(1): 139–44.

O'Sullivan AK, Thompson D, Drummond MF. Collection of health-economic data alongside clinical trials: Is there a future for piggyback evaluations? *Value Health* 2005; 8(1): 67–79.

Petrou S, Gray A. Economic evaluation alongside randomised controlled trials: Design, conduct, analysis, and reporting. *BMJ* 2011a; 342: d1548.

Petrou S, Gray A. Economic evaluation using decision analytical modelling: Design, conduct, analysis, and reporting. *BMJ* 2011b; 342: d1766.

Pocock SJ. *Clinical Trials: A Practical Approach.* Chichester: John Wiley, 1983.

Popat R, Khan I, Dickson J, Cheesman S, Smith D, D'Sa S, Rabin N, Yong K. An alternative dosing strategy of lenalidomide for patients with relapsed multiple myeloma. *Br. J. Haematol.* 2015; 168(1): 148–51.

Postma MJ, de Vries R, Welte R, Edmunds WJ. Health economic methodology illustrated with recent work on Chlamydia screening: The concept of extended dominance. *Sex. Transm. Infect.* 2008; 84(2): 152–4.

Ramsey S, Willke R, Briggs A, Brown R, Buxton M, Chawla A, Cook J, Glick H, Liljas B, Petitti D, Reed S. Good research practices for cost-effectiveness analysis alongside clinical trials: The ISPOR RCT-CEA Task Force report. *Value Health* 2005; 8(5): 521–33.

Raikou M, McGuire A. Estimating medical care costs under conditions of censoring. *J Health Econ.* 2004; 23(3): 443–70.

Ramsey SD, Willke RJ, Glick H, Reed SD, Augustovski F, Jonsson B, Briggs A, Sullivan SD. Cost-effectiveness analysis alongside clinical trials II – An ISPOR Good Research Practices Task Force report. *Value Health* 2015; 18(2): 161–72.

Rascati KL. *Essentials of Pharmacoeconomics,* 2nd edn. Baltimore: Lippincott Williams & Wilkins, 2013.

Regan J, Yarnold J, Jones PW, Cooke NT. Palliation and life quality in lung cancer; how good are clinicians at judging treatment outcome? *Br J Cancer* 1991; 64(2): 396–400.

Rittenhouse BE, Dulisse B, Stinnett AA. At what price significance? The effect of price estimates on statistical inference in economic evaluation. *Health Econ.* 1999; 8(3): 213–19.

Robins JM, Finkelstein DM. Correcting for noncompliance and dependent censoring in an AIDS Clinical Trial with inverse probability of censoring weighted (IPCW) log-rank tests. *Biometrics* 2000; 56(3): 779–88.

Robins JM, Hernán MA, Brumback B. Marginal structural models and causal inference in epidemiology. *Epidemiology* 2000; 11(5): 550–60.

Robinson R. Economic evaluation and health care. What does it mean? *BMJ* 1993; 307(6905): 670–3.

Rowen D, Brazier J, Young T, Gaugris S, Craig BM, King MT, Velikova G. Deriving a preference-based measure for cancer using the EORTC QLQ-C30. *Value Health* 2011; 14(5): 721–31.

Royston P, Lambert PC. *Flexible Parametric Survival Analysis using STATA: Beyond the Cox Model.* Texas: Stata Press, 2011.

Royston P, Parmar MK. Flexible parametric proportional-hazards and proportional-odds models for censored survival data, with application to prognostic modelling and estimation of treatment effects. *Stat. Med.* 2002; 21(15): 2175–97.

Rubin DB. *Multiple Imputation for Nonresponse in Surveys.* New York: John Wiley, 1987.

Rutten-van Mölken MP, van Doorslaer EK, van Vliet RC. Statistical analysis of cost outcomes in a randomized controlled clinical trial. *Health Econ.* 1994; 3(5): 333–45.

Sakia RM. The Box-Cox transformation technique: A review. *J. Roy. Stat. Soc. Series D (The Statistician)* 1992; 41(2): 169–78.

Santerre RE, Neun SP. *Health Economics: Theory, Insights and Industry Studies*, 5th edn. Orlando, FL: Cengage Learning, 2009.

Sarker SJ, Whitehead A. Factors affecting the sample size of cost-effectiveness trials. Technical Report, MPS Research Unit, Reading University, 2004.

Sarker SJ, Whitehead A, Khan I. A C++ program to calculate sample sizes for cost-effectiveness trials in a Bayesian framework. *Comput Methods Programs Biomed* 2013; 110(3): 471–89.

Sculpher MJ, Claxton K, Drummond M, McCabe C. Whither trial-based economic evaluation for health care decision making? *Health Econ.* 2006; 15(7): 677–87.

Senn S. *Statistical Issues in Drug Development.* New York: John Wiley, 1997.

Sharples LD, Edmunds J, Bilton D, Hollingworth W, Caine N, Keogan M, Exley A. A randomised controlled crossover trial of nurse practitioner versus doctor led outpatient care in a bronchiectasis clinic. *Thorax* 2002; 57(8): 661–6.

Shepherd FA, Pereira JR, Ciuleanu T, Tan EH, Hirsh V, Thongprasert S, Campos D, et al. National Cancer Institute of Canada Clinical Trials Group. Erlotinib in previously treated non-small-cell lung cancer. *N. Engl. J. Med.* 2005; 353(2): 123–32.

Siani C, Moatti J-P. The handling of uncertainty in economic evaluations of health care strategies. *Rev. Epidemiol. Sante Publique* 2003; 51(2): 255–76.

Simons CL, Rivero-Arias O, Yu LM, Simon J. Multiple imputation to deal with missing EQ-5D-3L data: Should we impute individual domains or the actual index? *Qual. Life Res.* 2015; 24(4): 805–15.

Sleeper LA, Harrington DP. Regression splines in the Cox model with application to covariate effects in liver disease. *J. Am. Stat. Assoc.* 1990; 85: 941–9.

Slevin ML, Plant H, Lynch D, Drinkwater J, Gregory WM. Who should measure quality of life, the doctor or the patient? *Br. J. Cancer* 1988; 57(1): 109–12.

Soares MO, Iglesias CP, Bland JM, Cullum N, Dumville JC, Nelson EA, Torgerson DJ, Worthy G: VenUS II team. Cost effectiveness analysis of larval therapy for leg ulcers. *BMJ* 2009; 338: b825.

Spiegelhalter DJ, Abrams KR, Myles JP. *Bayesian Approaches to Clinical Trials and Health Care Evaluation.* Chichester: John Wiley, 2004.

Stadtmauer EA, Weber DM, Niesvizky R, Belch A, Prince MH, San Miguel JF, Facon T, et al. Lenalidomide in combination with dexamethasone at first relapse in comparison with its use as later salvage therapy in relapsed or refractory multiple myeloma. *Eur. J. Haematol.* 2009; 82(6): 426–32.

Stinnett AA, Mullahy J. Net health benefits: A new framework for the analysis of uncertainty in cost-effectiveness analysis. *Med. Decis. Making* 1998; 18(Suppl.): S68–80.

Stokes ME, Davis CS. *Categorical Data Analysis Using the SAS System.* Cary: SAS Institute Publication, 2009.

Thatcher N, Chang A, Parikh P, Pereira JR, Ciuleano T, Pawel JV, Thongprasert S, et al. Geftinib plus best supportive care in previously treated patients with refractory advanced non-small-cell lung cancer: Results from a randomised, placebo-controlled, multicentre study (Iressa Survival Evaluation in Lung Cancer). *Lancet* 2005; 366(9496): 1527–37.

Thompson SG, Nixon RM, Grieve R. Addressing the issues that arise in analysing multicentre cost data, with application to a multinational study. *J. Health Econ.* 2006; 25(6): 1015–28.

Vale CD, Maurelli VA. Simulating multivariate nonnormal distributions. *Psychometrika.* 1983; 48: 465–71.

Welton N, Ades AE. Research decisions in the face of heterogeneity: What can a new study tell us? *Health Econ.* 2012; 21(10): 1196–200.

Whitehead J, Valdés-Márquez E, Johnson P, Graham G. Bayesian sample size for exploratory clinical trials incorporating historical data. *Stat Med.* 2008; 27: 2307–27.

Wijeysundera HC, Wang X, Tomlinson G, Ko DT, Krahn MD. Techniques for estimating health care costs with censored data: An overview for the health services researcher. *Clinicoecon. Outcomes Res.* 2012; 4: 145–55.

Willan AR, Lin DY, Cook RJ, Chen EB. Using inverse-weighting in cost-effectiveness analysis with censored data. *Stat. Methods Med. Res.* 2002; 11(6): 539–51.

Willke RJ, Glick HA, Polsky D, Schulman K. Estimating country-specific cost-effectiveness from multinational clinical trials. *Health Econ.* 1998; 7(6): 481–93.

Zhang L, Ma S, Song X, Han B, Cheng Y, Huang C, Yang S, Liu X, Liu Y, Lu S, Wang J, Zhang S, Zhou C, Zhang X, Hayashi N, Wang M; INFORM investigators. Gefitinib versus placebo as maintenance therapy in patients with locally advanced or metastatic non-small-cell lung cancer (INFORM; C-TONG 0804): A multicentre, double-blind randomised phase 3 trial. *Lancet Oncol.* 2012; 13(5): 466–75.

Bibliography

Al MJ, Van Hout BA. A Bayesian approach to economic analyses of clinical trials: The case of stenting versus balloon angioplasty. *Health Econ.* 2000; 9(7): 599–609.

Arnold RJG. *Pharmacoeconomics: From Theory to Practice.* Boca Raton: CRC Press, 2010.

Arnold RG, Kotsanos JG. Panel: Methodological issues in conducting pharmacoeconomic evaluations – retrospective and claims database studies. *Value Health* 1999; 2(2): 82–7.

Askew RL, Swartz RJ, Xing Y, Cantor SB, Ross MI, Gershenwald JE, Palmer JL, Lee JE, Cormier JN. Mapping FACT-melanoma quality-of-life scores to EQ-5D health utility weights. *Value Health* 2011; 14(6): 900–6.

Bennett CL, Westerman IL. Economic analysis during phase III clinical trials: Who, what, when, where, and why? *Oncology (Williston Park)* 1995; 9(11 Suppl.): 169–75.

Berlin JA, Cepeda MS. Some methodological points to consider when performing systematic reviews in comparative effectiveness research. *Clin. Trials* 2012; 9(1): 27–34.

Bleichrodt H, Filko M. New tests of QALYs when health varies over time. *J. Health Econ.* 2008; 27(5): 1237–49.

Blough DK, Ramsey S, Sullivan SD, Yusen R. Nett Research Group. The impact of using different imputation methods for missing quality of life scores on the estimation of the cost-effectiveness of lung-volume-reduction surgery. *Health Econ.* 2009; 18(1): 91–101.

Brazier J, Ratcliffe J, Salomon JA, Tsuchiya A. *Measuring and Valuing Health Benefits for Economic Evaluation.* Oxford: Oxford University Press, 2007.

Brennan A, Chick SE, Davies R. A taxonomy of model structures for economic evaluation of health technologies. *Health Econ.* 2006; 15(12): 1295–1310.

Briggs A, Fenn P. Confidence intervals or surfaces? Uncertainty on the cost-effectiveness plane. *Health Econ.* 1998; 7(8): 723–40.

Briggs A, Clark T, Wolstenholme J, Clarke P. Missing … presumed at random: Cost-analysis of incomplete data. *Health Econ.* 2003; 12: 377–92.

Bryan S, Williams I, McIver S. Seeing the NICE side of cost-effectiveness analysis: A qualitative investigation of the use of CEA in NICE technology appraisals. *Health Econ.* 2007; 16(2): 179–93.

Buxton MJ, Drummond MF, Van Hout BA, Prince RL, Sheldon TA, Szucs T, Vray M. Modelling in economic evaluation: An unavoidable fact of life. *Health Econ.* 1997; 6(3): 217–27.

Cancer Research UK GEDDIS Trial unpublished results: http://www.isrctn.com/ISRCTN07742377.

Cantoni E, Ronchetti E. A robust approach for skewed and heavy-tailed outcomes in the analysis of health care expenditures. *J. Health Econ.* 2006; 25(2): 198–213.

Claxton K. Heterogeneity in cost-effectiveness of medical interventions: The Cochrane Review 2008; challenges of matching patients to appropriate care; 3 November 2011; 14th Annual ISPOR Congress, 1–8 November 2011.

Contreras-Hernández I, Mould-Quevedo JF, Torres-González R, Goycochea-Robles MV, Pacheco-Domínguez RL, Sánchez-García S, Mejía-Aranguré JM, Garduño-Espinosa J. Cost-effectiveness analysis for joint pain treatment in patients with osteoarthritis treated at the Instituto Mexicano del Seguro Social (IMSS): Comparison of nonsteroidal anti-inflammatory drugs (NSAIDs) vs. cyclooxy-genase-2 selective inhibitors. *Cost Eff. Resour. Alloc.* 2008; 12(6): 21.

Cooper NJ, Sutton AJ, Abrams KR, Turner D, Wailoo A. Comprehensive decision analytical modelling in economic evaluation: A Bayesian approach. *Health Econ.* 2004; 13(3): 203–26.

Coyle D, Drummond MF. Analyzing differences in the costs of treatment across centers within economic evaluations. *Int. J. Technol. Assess. Health Care* 2001; 17(2): 155–63.

Craig BA, Sendi PP. Estimation of the transition matrix of a discrete-time Markov chain. *Health Econ.* 2002; 11(1): 33–42.

Craig MB. The duration effect: A link between TTO and VAS values. *Health Econ.* 2009; 18(2): 217–25.

Cramer JA, Spilker B. *Quality of Life and Pharmacoeconomics.* Philadelphia: Lippincott-Raven, 1998.

Díaz-Ordaz K, Kenward MG, Grieve R. Handling missing values in cost effectiveness analyses that use data from cluster randomized trials. *J. Roy Stat Soc Ser A* (Statistics in Society) 2014; 177(2): 457–74.

Dreyer NA, Schneeweiss S, McNeil BJ, Berger ML, Walker AM, Ollendorf DA, Gliklich RE. GRACE principles: Recognizing high-quality observational studies of comparative effectiveness. GRACE Initiative. *Am. J. Manag. Care* 2010; 16(6): 467–71.

Eckermann S, Willan AR. Globally optimal trial design for local decision making. *Health Econ.* 2009; 18(2): 203–16.

European Medicines Agency. Guideline on the missing data; EMA/CPMP/EWP/1776/99 Rev.

European Medicines Agency; Oncology Working Party. Reflection Paper on the use of patient reported outcome (PRO) measures in oncology studies. EMA/CHMP/292464/2014. EMA, 2014.

Evidence Review Group. The clinical- and cost-effectiveness of lenalidomide for multiple myeloma in people who have received at least one prior therapy: An evidence review of the submission from CELGENE; PenTAG on behalf of NICE, 1 September 2008.

Feenstra TL, van Baal PH, Gandjour A, Brouwer WB. Future costs in economic evaluation. A comment on Lee. *J. Health. Econ.* 2008; 27(6): 1645–9; discussion 1650–1.

Fox-Rushby J, Cairns J. eds., *Economic Evaluation.* Maidenhead: Open University Press, 2005.

Gardiner JC. Modeling heavy-tailed distributions in healthcare utilization by parametric and Bayesian methods. *SAS Global Forum,* 2012.

Graves N, Walker D, Raine R, Hutchings A, Roberts JA. Cost data for individual patients included in clinical studies: No amount of statistical analysis can compensate for inadequate costing methods. *Health Econ.* 2002; 11(8): 735–9. Review.

Grieve R, Cairns J, Thompson SG. Improving costing methods in multicentre economic evaluation: The use of multiple imputation for unit costs. *Health Econ.* 2010; 19(8): 939–54.

Hardy KJ, Szczepura A, Davies R, Bradbury A, Stallard N, Gossain S, Walley P, Hawkey PM. A study of the efficacy and cost-effectiveness of MRSA screening and monitoring on surgical wards using a new, rapid molecular test (EMMS) BMC. *Health Serv. Res.* 2007; 7: 160.

Hawkins N, Scott DA. Reimbursement and value-based pricing: Stratified cost-effectiveness analysis may not be the last word. *Health Econ.* 2011; 20(6): 688–98.

Heitjan DF, Li H. Bayesian estimation of cost-effectiveness: An importance-sampling approach. *Health Econ.* 2004; 13(2): 191–8.

Hlatky MA. Economic endpoints in clinical trials. *Epidemiol Rev.* 2002; 24(1): 80–4.

Jackson CH, Thompson SG, Sharples LD. Accounting for uncertainty in health economic decision models by using model averaging. *J. R. Stat. Soc. Ser. A Stat. Soc.* 2009; 172(2): 383–404.

Jackson CH, Bojke L, Thompson SG, Claxton K, Sharples LD. A framework for addressing structural uncertainty in decision models. *Med. Decis. Making* 2011; 31(4): 662–74.

Jonas ED, Wilkins MT, Bangdiwala, Bann CM, Morgan LC, Thaler KJ, Amick HR, Gartlehner G. Findings of a Bayesian mixed treatment comparison meta analyses: Comparison and exploration using real-world trial data and simulation. Agency for Healthcare Research and Quality; February 2013.

Jönsson L, Sandin R, Ekman M, Ramsberg J, Charbonneau C, Huang X, Jönsson B, Weinstein MC, Drummond M. Analyzing overall survival in randomized controlled trials with crossover and implications for economic evaluation. *Value Health* 2014; 17(6): 707–13.

Kaambwa B, Billingham L, Bryan S. Mapping utility scores from the Barthel index. *Eur. J. Health Econ.* 2013; 14(2): 231–41.

Karnon J, Goyder E, Tappenden P, McPhie S, Towers I, Brazier J, Madan J. A review and critique of modelling in prioritising and designing screening programmes. *Health Technol. Assess.* 2007; 11(52): iii–iv, ix–xi, 1–145.

Klein JP, Moeschberger ML. *Survival Analysis: Techniques for Censored and Truncated Data (Statistics for Biology and Health)*, 3rd edn. Springer, 2005.

Kreif N, Grieve R, Radice R, Sadique Z, Ramsahai R, Sekhon JS. Methods for estimating subgroup effects in cost-effectiveness analyses that use observational data. *Med. Decis. Making* 2012; 32(6): 750–63.

Kunz R, Vist G, Oxman AD. Randomisation to protect against selection bias in healthcare trials. *Cochrane Database Syst Rev.* 2007; (2): MR000012.

Latimer N, Abrams KR. NICE DSU Technical Support Document 14: Survival Analysis for Economic Evaluations alongside Clinical Trials – Extrapolation with Patient Level Data; Report by the Decision Support Unit, June 2011.

Longworth L, Rowen D. NICE DSU Technical Support Document 10: The Use of Mapping Methods to Estimate Health Utility State Values; Report by the Decision Support Unit, April 2011.

Manca A, Sculpher MJ, Goeree R. The analysis of multinational cost-effectiveness data for reimbursement decisions. *Pharmacoeconomics* 2010; 28(12): 1079–96.

Marsh K, Phillips CJ, Fordham R, Bertranou E, Hale J. Estimating cost-effectiveness in public health: A summary of modelling and valuation methods. *Health Econ. Rev.* 2012; 2(1): 17.

Muening P. *Cost-Effectiveness Analysis in Health. A Practical Approach*, 2nd edn. New York: John Wiley, 2008.

Nixon RM, Thompson SG. Methods for incorporating covariate adjustment, subgroup analysis and between-centre differences into cost-effectiveness evaluations. *Health Econ.* 2005; 14(12): 1217–29.

Oostenbrink JB, Al MJ. The analysis of incomplete cost data due to dropout. *Health Econ.* 2005; 14(8): 763–76.

Papaioannou D, Brazier J, Paisley S. NICE DSU Technical Support Document 9: The Identification Review and Synthesis of Health State Utility Values from the Literature; Report by the Decision Support Unit, October 2010.

Parker M, Haycox A, Graves J. Estimating the relationship between preference-based generic utility instruments and disease-specific quality-of-life measures in severe chronic constipation: Challenges in practice. *Pharmacoeconomics* 2011; 29(8): 719–30.

Philips Z, Ginnelly L, Sculpher M, Claxton K, Golder S, Riemsma R, Woolacoot N, Glanville J. Review of guidelines for good practice in decision-analytic modelling in health technology assessment. *Health Technol. Assess.* 2004; 8(36): iii–iv, ix–xi, 1–158.

Pohlmeier W, Ulrich V. An econometric model of the two-part decision making process in the demand for health care. *J. Hum. Resour.* 1995; 30(2): 339–61.

Prieto L, Sacristán JA. Problems and solutions in calculating quality-adjusted life years (QALYs). *Health Qual. Life. Outcomes* 2003; 1: 80.

Ridderstråle M, Jensen MM, Gjesing RP, Niskanen L. Cost-effectiveness of insulin detemir compared with NPH insulin in people with type 2 diabetes in Denmark, Finland, Norway, and Sweden. *J. Med. Econ.* 2013; 16(4): 468–78.

Rychlik R. *Strategies in Pharmacoeconomics and Outcomes Research.* New York: Pharmaceuticals Products Press, 2002.

Sassi F. Calculating QALYs, comparing QALY and DALY calculations. *Health Policy Plan.* 2006; 21(5): 402–8.

Schmitz S, Adams R, Walsh CD, Barry M, FitzGerald O. A mixed treatment comparison of the efficacy of anti-TNF agents in rheumatoid arthritis for methotrexate non-responders demonstrates differences between treatments: A Bayesian approach. *Ann. Rheum. Dis.* 2012; 71(2): 225–30.

Scott DA, Boye KS, Timlin L, Clark JF, Best JH. A network meta-analysis to compare glycaemic control in patients with type 2 diabetes treated with exenatide once weekly or liraglutide once daily in comparison with insulin glargine, exenatide twice daily or placebo. *Diabetes Obes. Metab.* 2013; 15(3): 213–23.

Sculpher M. Reflecting heterogeneity in patient benefits: The role of subgroup analysis with comparative effectiveness. *Value Health* 2010; 13(Suppl. 1): S18–S21.

Severens JL, Brunenberg DE, Fenwick EA, O'Brien B, Joore MA. Cost-effectiveness acceptability curves and a reluctance to lose. *Pharmacoeconomics* 2005; 23(12): 1207–14. Review.

Simes RJ, Coates AS. Patient preferences for adjuvant chemotherapy of early breast cancer: How much benefit is needed? *J. Natl. Cancer Inst. Monogr.* 2001; (30): 146–52.

Singer DJ. Using SAS PROC MIXED to fit multilevel models, hierarchical models, and individual growth models. *J. Educ. Behav. Stat.* 1988; (4): 323–55.

Soto J. Health economic evaluations using decision analytic modeling. *Int. J. Technol. Assess. Health Care* 2002; 18(1): 94–111.

Sutton A, Ades AE, Cooper N, Abrams K. Use of indirect and mixed treatment comparisons for technology assessment. *Pharmacoeconomics* 2008; 26(9): 753–67. Review.

Tambour M, Zethraeus N, Johannesson M. A note on confidence intervals in cost-effectiveness analysis. *Int. J. Technol. Assess. Health Care* 1998; 4: 467–71.

Torrance GW. Measurement of health state utilities for economic appraisal. *J. Health Econ.* 1986; 5(1): 1–30.

Tsuchiya A. The allocation of health care resources: The basic concepts (in Japanese). In Kato H, Kamo N, eds. *Handbook for Bioethics*, Tokyo: Sekai Shiso Sha, 1998.

Tsuchiya A, Williams A. Welfare economics and economic evaluation. In Drummond M, McGuire A, eds., *Economic Evaluation in Health Care: Theory and Practice*, Chapter 2. New York: Oxford University Press, 2001.

Watkins C, Huang X, Latimer N, Tang Y, Wright EJ. Adjusting overall survival for treatment switches: Commonly used methods and practical application. *Pharm. Stat.* 2013; 12(6): 348–57.

Willan AR. Sample size determination for cost-effectiveness trials. *Pharmacoeconomics* 2011; 29(11): 933–49.

Willan AR, Briggs AH. *Statistical Analysis of Cost-Effectiveness Data.* Chichester: John Wiley, 2006.

Willan AR, Eckermann S. Accounting for between-study variation in incremental net benefit in value of information methodology. *Health Econ.* 2012; 21(10): 1183–95.

Willan AR, Kowgier ME. Cost-effectiveness analysis of a multinational RCT with a binary measure of effectiveness and an interacting covariate. *Health Econ.* 2008; 17(7): 777–91.

Willan AR, O'Brien BJ. *Health Econ.* 1996a; 5(4): 297–305. Erratum in *Health Econ.* 1999; 8(6): 559.

Willan AR, O'Brien BJ. Confidence intervals for cost-effectiveness ratios: An application of Fieller's theorem. *Health Econ.* 1996b; 5: 297–305.

Willan AR, Briggs AH, Hoch JS. Regression methods for covariate adjustment and subgroup analysis for non-censored cost-effectiveness data. *Health Econ.* 2004; 13(5): 461–75.

Wordsworth S, Ludbrook A. Comparing costing results in across country economic evaluations: The use of technology specific purchasing power parities. *Health Econ.* 2005; 14(1): 93–9.

Yang X, Shoptaw S, Nie K, Liu J, Belin TR. Markov transition models for binary repeated measures with ignorable and nonignorable missing values. *Stat. Methods Med. Res.* 2007; 16(4): 347–64.

Zhao H, Bang H, Wang H, Pfeifer PE. On the equivalence of some medical cost estimators with censored data. *Stat. Med.* 2007; 26(24): 4520–30.

Zhao H, Zuo C, Chen S, Bang H. Nonparametric inference for median costs with censored data. *Biometrics* 2012; 68(3): 717–25.

Index